THE COMPLETE BOOK OF TRADITIONAL GUERNSEY AND JERSEY KNITTING

Yoked gansey with unusual cable pattern, seen in Whitby *Sutcliffe Gallery, Whitby*

The Critic, photographed by Frank Sutcliffe in Whitby about 1906, shows a group of gansey-clad fishermen pointing out the important points of lobster-pot making to a young fisherman. Their ganseys show the simple patterns in use at that time. The man sitting on a lobster pot is also wearing long hand-knitted Abb wool stockings inside his thigh length sea-boots, and his gansey shows the split borders often used in place of a ribbed welt.
Photograph by permission of the Sutcliffe Gallery, Whitby.

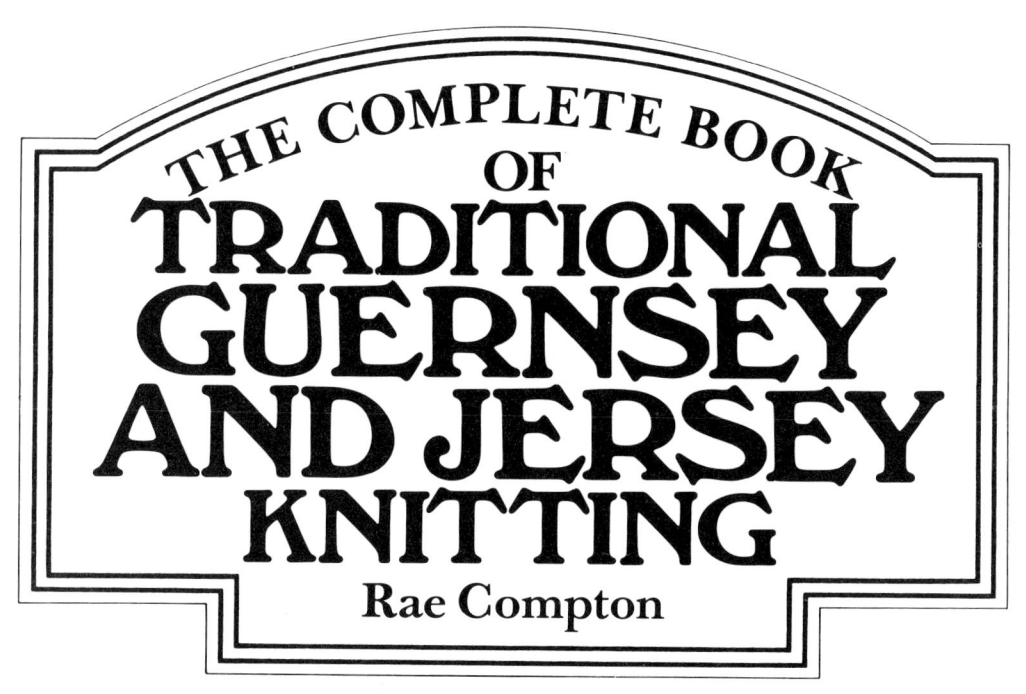

THE COMPLETE BOOK OF TRADITIONAL GUERNSEY AND JERSEY KNITTING
Rae Compton

B.T. BATSFORD LTD   LONDON

© Rae Compton 1985
First published 1985

All rights reserved. No part of this publication may be reproduced, in any form or by any means, without permission from the Publisher

ISBN 0 7134 4125 9

Typeset by Servis Filmsetting Ltd, Manchester
and printed in Great Britain by
Anchor-Brendon Ltd
Tiptree, Essex
for the publishers
B.T. Batsford Ltd
4 Fitzhardinge Street
London W1H 0AH

Diagrams by Vivienne Compton

Contents

Acknowledgements 6

1 Introduction to guernseys 8
2 Research problems 13
3 Guernsey basics 20
4 Cornwall and Devon 31
5 Norfolk 41
6 Yorkshire and Northern England 52
7 South and East Scotland 69
8 North and West Scotland 86
9 Designing a guernsey 98
10 Guernsey designs 104

Bibliography 138
Conversion charts 139
List of Suppliers 141
Index 142

ACKNOWLEDGEMENTS

My thanks must go to everyone who helped in any way to piece together the story of the guernseys, past and present. The list is too long to mention every individual, but each person was important to the whole and I am indeed grateful for all their help.

In particular I must thank my daughter Vivienne who repeatedly discarded carefully worked charts as new information led to necessary changes. My thanks also go to the following, the numbers in brackets referring to photographs kindly lent: Mary Wright, Andrew Lanyon, Sheila de Burlet, Mr Douch and Mr Penhallurick of the Royal Institution of Cornwall, Truro (31, 32, 33, 34, 36, 44, 46, 158); Martin Warren of Cromer Museum and Norfolk Museum Services (5, 50, 51, 56, 59, 60, 63, 65); Michael Harvey of London (47); Mr Crimlisk of Filey (85, 87, 93); Mrs Scales of Flamborough Head and the Walkington family of Bridlington (79, 82, 149); Mr J. Robin Lidster of Scarborough (1, 69, 96, 97, 109, 145) and Scarborough Museum and Library Services (146); Eyemouth Museum (102); Mr Michael Fairnie, Mrs Clark, Mrs Brown and Mrs Orr of Fisherrow and Musselburgh Library; the Scottish Fisheries Museum, Anstruther (100, 106, 108, 111, 113, 116), and Miss Mary Murray of Anstruther; St Andrews University Library; Dundee, Arbroath and Peterhead Museum staff; Mr James Slater and Mrs Sutherland, Portsoy; Henrietta Munro and the fisher people of Thurso, Caithness; Mary More of Hopeman and Mr and Mrs Stewart (125) and Mrs Stewart senior of Lossiemouth; Mr Donald Ross and Mrs Ross of Cromarty; Richard Poppleton and Sons, Wakefield, for yarn and for their help and interest; Sutcliffe Gallery, Whitby (half title, frontispiece, 3); Doris Cleverly, Helmsley (2, 72); National Museum of Antiquities, Edinburgh (4); R.N.L.I. (17); Cyril Noall, St Ives (25); B.T. Batsford Ltd (54, 161); Mr and Mrs Mainprize, Bridlington (74, 76, 77); Kingston upon Hull City Museums & Art Galleries (78); Scottish Publications Ltd, Edinburgh (103); Arbuthnot Museum, Peterhead (119); Miss Donella Mackay, Thurso (132, 163); Mrs Ralston, Campbeltown (152) and Miss Stewart, Edinburgh (153), Rev. Peter Longridge, Appledore.

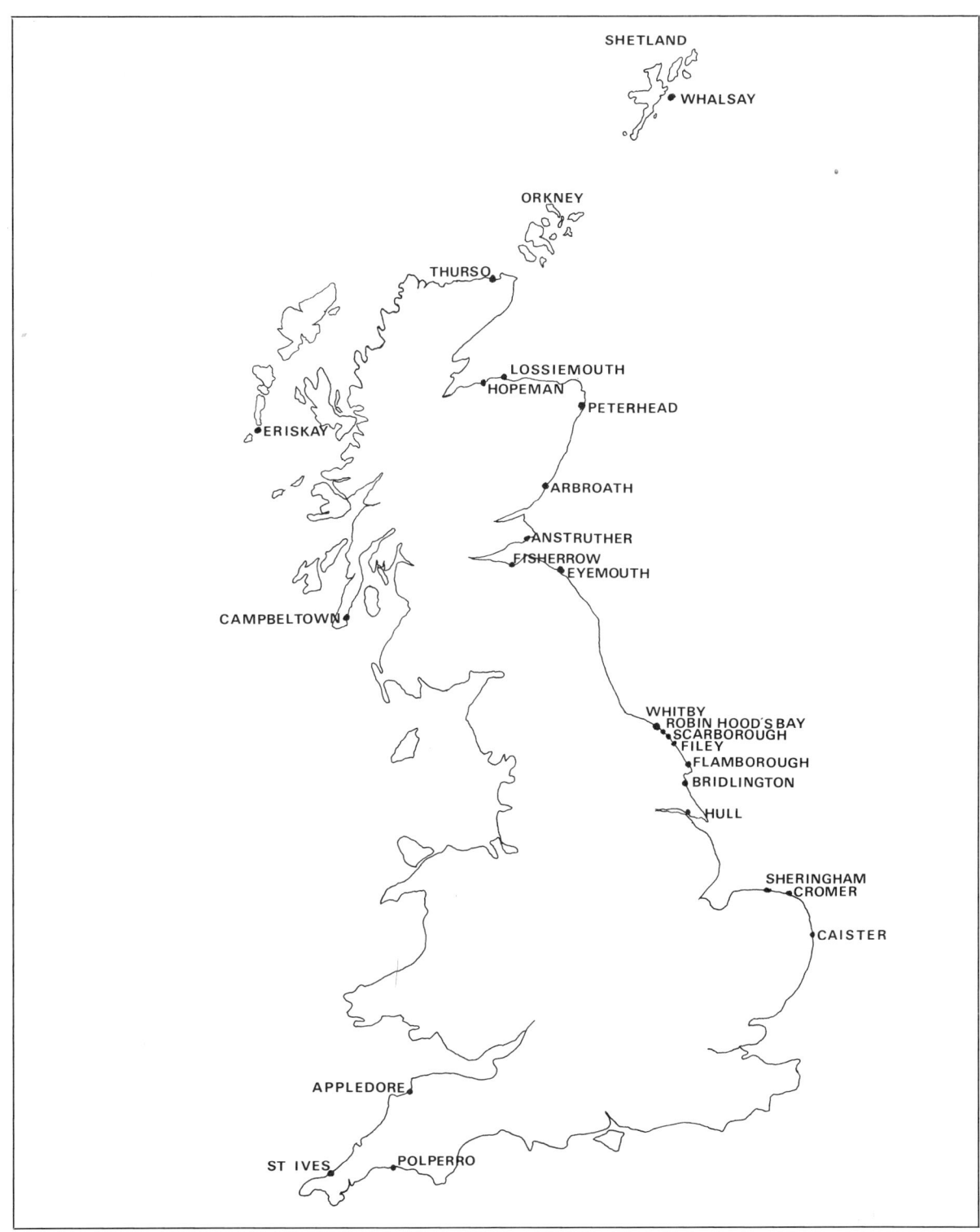

Map of British Isles showing places where specific designs were found

1 Introduction to guernseys

Raised on the east coast of Scotland, there never seems to have been a time when I was not aware of the dark blue, almost black guernseys worn by the fishermen in the tiny harbours that were my playground. It was not until I was confined to a London office, no longer free to chase the sun-drenched spray along the old stone walls, that I realised the wealth of pattern in their stitches or the nation-wide story of which they were only a part.

It was then that I resolved to return, not to watch the breaking waves or delight in the popping of dried seaweed blisters, not even to watch the boats returning to harbour, each with its creamy petticoat of wake and lace-like frill of attendant gulls, but to seek out the knitters and ask them about the patterns their stitches formed. It was not until this book was planned, however, that I was able to make my return a reality.

Like many knitters before me, I had learned something of guernsey knitting from *Guernsey and Jersey Patterns* by Gladys Thompson (first published by Batsford in 1955), and it was with pleasure that I set out to trace her footsteps. The pleasure was tinged with concern, however, that time had passed too quickly and that perhaps the old tradition of guernsey knitting had been swept aside by the age of television and superstores.

I did not mind if the donkeys no longer carried young visitors down the cobbled streets of Robin Hood's Bay with guernsey-clad youths in attendance (1), but I did want to find that the higgledy-piggledy lanes still existed, and that when the sun shone on

1 *Friendly transport for a young visitor*

sleeping cats and flowering geraniums, a housewife or two would still be knitting by the kitchen door until the men came home and the dinner was cooked (2).

Indeed the world had changed and perhaps changed again in those years, but although every clothes-line was not filled with drying guernseys on even dry Monday mornings, cats still found sunny corners and knitters still knitted, in many places even with the same yarn.

This book contains some of what I found, and I hope that it will help you to enjoy making garments that will be useful, win admiration and, in turn, be copied by others. In all the many conversations that have led to this book, only once have I listened and admired for a whole afternoon only to be asked not to print a word of what had been said. From north to south I have met with interest and help to a degree that I did not expect, and this heritage of patterns comes with the hope that the experiments and treasures of the past will find a new use in the future.

To include the maximum number of stitch patterns, photographs of swatches have been omitted and charts have been provided instead. Photographs, which might have shown some of the old patterns more clearly, would have required many words to accompany them, whereas the charts can be used in two ways. For working in rounds, each chart shows the pattern as it appears on the right side of the knitting. Each purl stitch is shown with an x and each knit stitch with a blank square. Each round will read the chart from right to left, and in most cases the pattern repeat will be marked or will be obvious. For working in rows, you must adapt wrong side rows so that they produce what you see on the right side on the chart. This means that a stitch shown by an x must be knitted to give the purl effect on the right side, and stitches which appear on the right side as knit stitches must be worked as purl stitches.

2 *Knitting in the sun, Staithes*

ABBREVIATIONS

The standard knitting abbreviations are used throughout and are listed here. Where any stitch pattern or design uses an unusual abbreviation or a cable or rope abbreviation, it will be printed before the start of the instructions concerned.

K	knit
P	purl
st(s)	stitch(es)
st st	stocking stitch
dec	decrease
inc	increase
m1	make 1 by lifting thread before next st and working into back of it
alt	alternate
beg	beginning
cont	continue
patt	pattern
psso	pass slipped st over
rep	repeat
rem	remaining
tbl	through back of loop(s)
tog	together
sl 1	slip 1
yon	yarn over needle
in	inch(es)
cm	centimetre(s)
CN	cable needle
g	gramme

2 Research problems

It is evidently comparatively easy to be read and quoted by everyone, year in year out, even if you write without verification and on a subject on which you have very little knowledge. On the other hand, it seems to be endlessly difficult to correct any of this wrongly reported material. In Caithness today it is still believed by some that guernseys were patterned for recognition in cases of drowning, particularly 'if they were drooned wi' their heid aff'. No knitter has ever confirmed that a pattern was designed for this purpose, although all knitters would recognise their own handiwork under any circumstance.

PATTERN NAMES

When any article on guernsey knitting is published, it sooner or later seems to include a list of lyrical names not always recognised by the knitters. Steps and ladders are often used, but did a fisherwoman really look at the steps up the cliff and hurry home to create them in her knitting? Blocks and slabs too have their patterns, but it seems doubtful if any knitter gazed on the rocks and was inspired to form a pattern from them.

There are names that have come into common usage through the need to identify the most used stitches: birds' e'en, cat's teeth and hailstones, for instance. If you have never heard of them, perhaps you know them all better by the name moss stitch, because that is what they are. The first comes from Flamborough Head, the second from Scotland and the third from Norfolk.

REGIONAL DIFFERENCES
It is also popular belief that each small area can be identified by its particular design, variations of which serve the various families, each with his distinguishing line or border. How very simple this would be, but also completely boring. Guernseys are created by knitters who well know how the idea of a slight alteration comes into their heads while their fingers are busy and which will find its place in the next garment to be knitted. Nothing is uniform and it is surprising, with the enormous variety that there is, that there are styles found in one area more than in another.

MARRIAGE
Marriage is another reason for research proving difficult. Many a fisherman marries from his own community but many also take wives from other communities, sometimes even outside fishing altogether. If she is already a knitter, she may retain her own patterns and so suddenly Cornish patterns are to be found in Yorkshire, or Scottish patterns in Sheringham.

TRAVEL
In trying to identify each person, the fisherman's roadway seems to have been forgotten. Not only was travel by sea his way of life, but fish had to be caught and therefore followed as they moved their grounds annually: north for the southerner and south for the northerner. The movements of the fishermen resulted in the travelling of patterns, and so again a pattern knitted in Cornwall, sold in Plymouth and resold in Yarmouth, might very well end up in Thurso.

For any family with girls travelling to gut and pack herring, new patterns were inevitable. Fingers busy at gutting turned to knitting when there was time to stop or when the next catch was due to arrive.

REGIONAL SIMILARITIES
A closer look at various designs shows that they could usually be found in most areas, although each had many variations.

One banded pattern is found north and south and can be seen worn by a fisherman photographed in

Whitby in the 1850s (3), in an illustration for a typical fisherman's dress of the nineteenth century from Newhaven near Edinburgh (4), and worn by Gilbert 'Leather' Rook at Sheringham in Norfolk early this century (5). In more recent times this design has been used by many fashion designers for a chunky knit with different textures between the bands. Another pattern found in most areas was published by Weldon in 1880, with the suggestion that ladies with time to spare might knit it for the men of the Seamen's Mission. It was unbanded and

3 *Banded pattern worn here by Harry Freeman of Whitby, c. 1880 Photo by Frank Sutcliffe*

4 *A Newhaven fisherman in customary dress about 1850*

6 *Single and double diamonds*

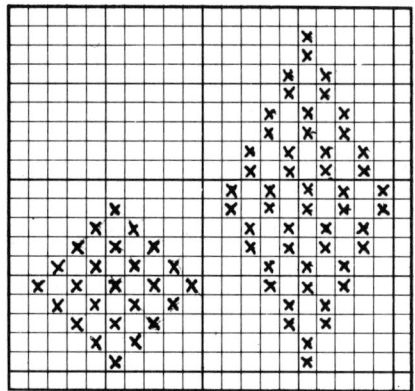

5 *Gilbert 'Leather' Rook of Sheringham c.1902 wearing an often seen banded pattern*

the entire yoke was worked in a small basket stitch similar to the background of John Cox's gansey (61) and often misnamed double moss stitch.

RECORDING PATTERNS

Every effort has been made to chart patterns correctly. This can be harder than it sounds and, when working from photographs, it is essential to take even greater care. Light, which plays an important part in any photograph, can distort what is actually in the photograph, making a rope pattern appear to recede when in fact it should stand out. Shadow on some smaller stitches, like basket stitch, can also make clear identification difficult.

Diamonds seem to confuse many people. A diamond worked in single moss, or when the pattern is alternated every row, will look less tall (6) than a diamond worked in double moss stitch taking twice as many rows. Small background patterns can also be distorted by light and need careful interpretation. Moss stitch and double moss stitch, marked 1 and 2 respectively on chart 7 are by no means the only background stitches used. The simplest variation is

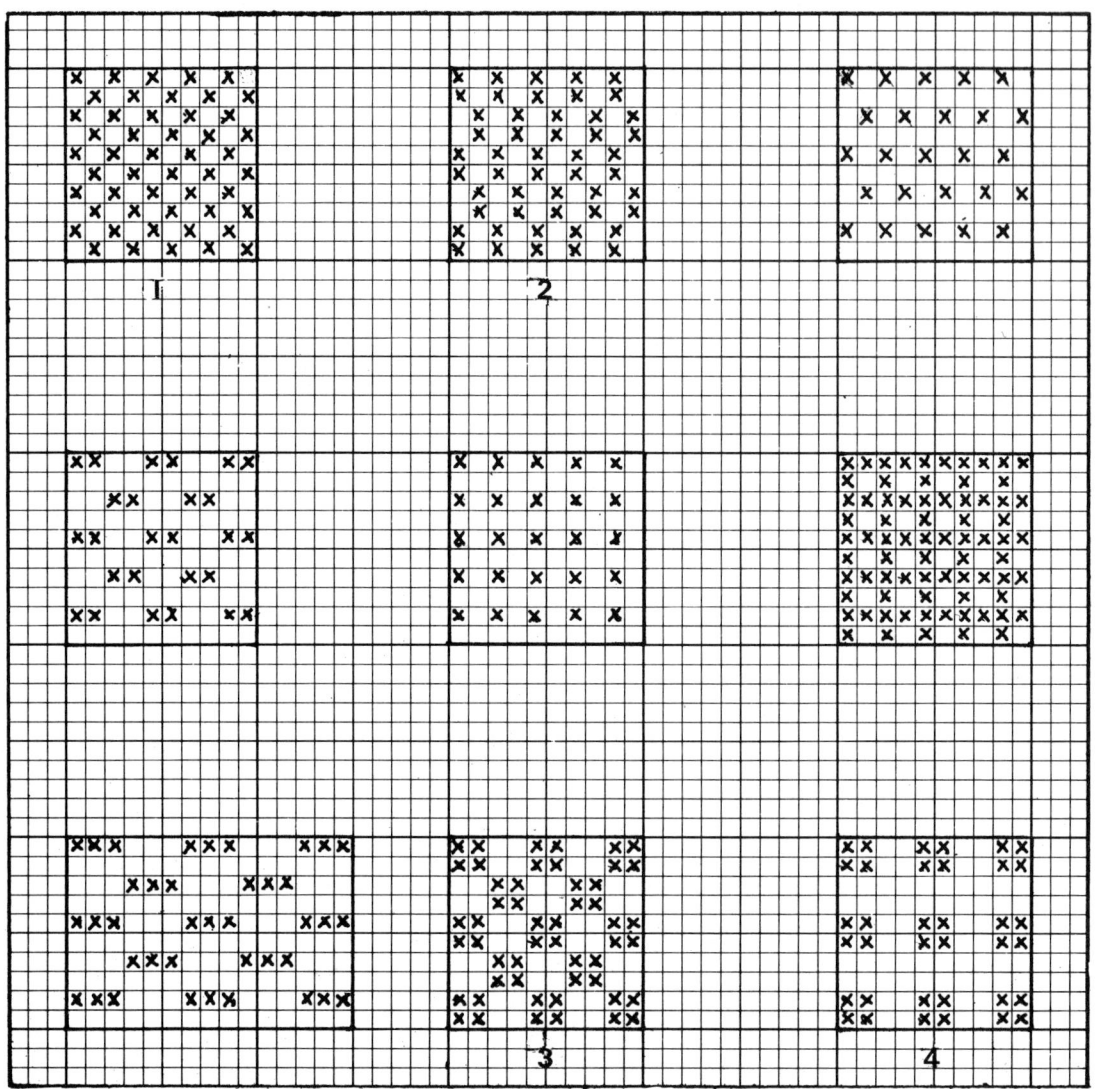

7 *Background or 'seeding' stitches*

to find both of these stitches expanded by having a plain row between the pattern rows. Other variations follow, including small grouped stitches like basket, box or Mary-Ann's stitch, marked 3, and 'Betty Martin', marked 4.

Cornish patterns can be most misleading. Here garter stitch seems to have been in great use, and even when ribs of single stitches are worked they tend to be a purl stitch only on alternate rows, giving only half the number of ridges that purling on every row would produce.

If you are trying to copy an old photograph, mark

out the pattern on graph paper. Once you have one motif drawn to your satisfaction, work the next to check that you have in fact done the correct number of rows.

DOUBTFUL ORIGINS

It is not surprising that it is always understood that guernseys, or ganseys as many called them, came originally from the Channel Islands as a later extension of their sock-knitting trade. This may well be the case, but it is exceedingly difficult to prove.

The Channel Island guernsey, which had ceased to be made prior to the last war, was made in the European way: round up to the neck, with sleeves worked independently to the top where the armhole was cut and the sleeve inserted. Only after the war was a different method introduced and guernsey knitting existed again, largely due to the efforts of the Women's Institute.

But in Britain, from possibly about the seventeenth century, two types of garments could have led more convincingly to guernseys. One of these was a fine knitted silk shirt or vest, often delicately patterned like brocade, with purl stitches on a stocking stitch background. The second, usually a child's garment, was a vest or undergarment knitted in cotton. Both garments were worked round, had seam stitches knitted in and included small areas of shaping so that the garment was not entirely shapeless. There seems to have been no area in Britain, at least in living memory, where the armholes were opened by cutting after knitting, except in Shetland where this method may still be used for the finely knitted Fair Isle garments.

3 Guernsey basics

For many years it has been said that the guernsey took its name from garments made in the Channel Islands when the knitting trade there was at its height. Those that put forward this claim called the fisherman's handknitted garment that was knitted without seams a guernsey, and gave the name jersey to garments made in finer wool, possibly knitted in sections and seamed together. There is little proof that this classification is correct. Throughout the British Isles the fisherman's garment is known in some places as a guernsey, in others as a gansey and in others it has always been and still is a jersey. The descriptive name of knit-frock has disappeared but there are still old knitters who can recall when it was known as a jersey-frock, which would seem to contradict any claim that it originated from the island of Guernsey.

Although the name may vary and the stitch patterns used can be endless, the basic garment is always knitted round on body and sleeves and in rows for the upper front and back, giving a seamless garment. To work in rounds, four or more needles were used, more often referred to as 'wires' than needles. Today they are available, neatly sized and ready packed, from wool shops and department stores, but in times gone by they were exactly as named, wires, cut from a length of steel wire and rounded to a comfortable point on the doorstep or other stone. Like many other commodities in the past, a knitter's wires were looked after, replaced only when necessary and handed on to daughter or grand-daughter when she no longer had use for them.

In guernsey knitting there is no room for tedious repetition. It seems to remain alive because, within the limits placed on it of working a seamless garment with neck and two sleeves, there is still a great deal of scope for personal expression and individuality. Generation after generation of knitters are still adding their own patterns, creating new textures, altering ways of placing designs and of shaping gussets or working shoulders and neckline.

Familiarity with the work does not breed contempt, but rather leaves knitters free to realise, while working one garment, that a small alteration on the next will make a new version, and that that in turn will lead on again to something else. There must still be guernseys lying forgotten in cupboards or drawers which would add to the following variations, but those in this book serve to show the basic differences to be found around the British coast.

CASTING ON
Many guernsey knitters claim that you must cast on double, but surprisingly this seems to have more meanings than are apparent on first consideration.

Casting on with two needles
Often called chain edge cast on, this is not worked with two strands of yarn but is used in Scotland where the word double indicates that it is not the single version of two needle casting on that is used. The single needle cast on, used mostly for lacy work, makes a new stitch by inserting the right-hand needle through the stitch on the left needle, and drawing through a loop of yarn. The double cast on, which makes a firm but elastic edge, is worked by slipping the right needle between two loops on the left needle (8), making a new stitch and drawing it through before placing it on the left needle (9).

Thumb method casting on
Again, this is not what springs to mind when double casting on is mentioned, but must be included because of the many knitters who consider it to be double because two ends of yarn are used.

A slip loop is made some way along the yarn, at a distance approximately three times the length of the work to be cast on. This loop is placed on the right needle, the yarn from the ball is held over the right

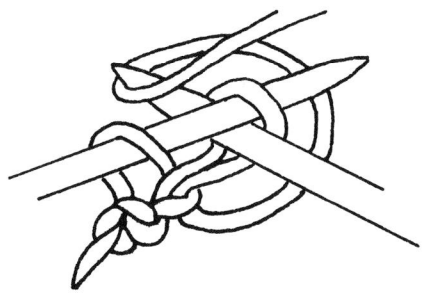

8 *Working two needle casting on*

9 *Two needle or chain edge casting on*

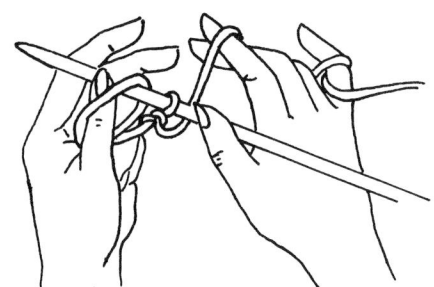

10 *Working thumb method of casting on*

11 *Thumb method or cable edge casting on*

fingers and the free length of yarn is formed into a loop by the left hand. Using the needle in the right hand, knit into the left-hand loop and make a stitch by bringing the right-hand yarn round the needle tip and through the left-hand loop (10). The new loop is then added to the right-hand needle and repeated until the correct number of loops is gained. The edge formed is strong and reasonably elastic, standing up to every-day wear (11). An alternative name for this stitch is cable casting on.

Three strand casting on

This is one of the strongest ways of working a reinforced edge which remains neat but is still elastic and more durable. It does become less attractive if the work is not on the fine needles that were always used to give a close texture and a firm, hard-wearing surface. It is worked in exactly the same way as the thumb method, but using two balls of yarn. In this way two strands are used to form the left-hand loops and are knitted through with two strands also. The double yarn is then used for the first four or six rows of rib before one strand is discarded and the garment is continued with one strand (12).

Knotted casting on

Seen to best advantage where a guernsey starts with a garter stitch border rather than a ribbed welt, this knotted edge is most effective.

Use the thumb method with a single strand of yarn. Work two stitches, then with the left needle tip

12 *Three strand casting on*

lift the first stitch on the right needle over the second stitch and off the needle tip, leaving one stitch. * Work two more stitches, again lifting the first over the second, leaving two stitches altogether on the right needle (13). Continue from * in this way until the required number of stitches is obtained (14).

Channel Island casting on
A method used in the Channel Islands seems to combine elements from the last two methods. The slip loop is made and placed on the right needle, leaving two strands hanging from it for the left-hand loops. Only one strand, however, is used on the right hand for making the actual stitch. * Wind the double end twice round the thumb and take the single yarn over the needle to form a strand across the needle, before knitting into the loop to make another stitch. This made stitch holds the strand over the needle in place and forms two stitches in the single movement (15). Repeat from * until the correct number of stitches is obtained. This will make an odd number, but if an even number is required, two stitches can be knitted together on the first row. This edging also sits most attractively on a plain bordered garment (16).

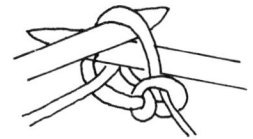

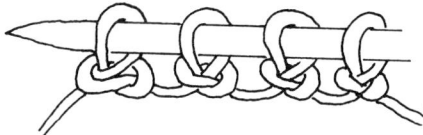

13 Lifting one stitch over the next for knotted edge casting on

14 Knotted edge casting on

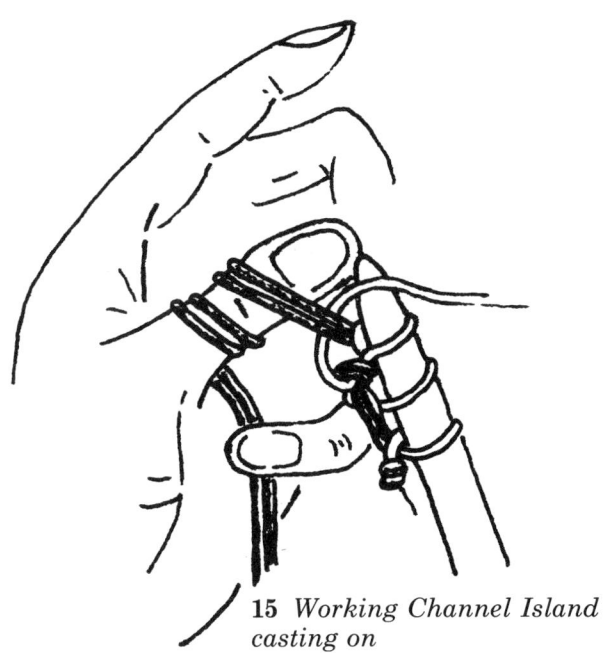

15 Working Channel Island casting on

16 Channel Island casting on

WELTS

There are two ways of beginning the lower edge of a guernsey: a border made in a flat stitch (17) or a ribbed welt (18).

Plain borders

A plain border, whether in garter stitch, moss stitch or in a ridged pattern, is usually worked in two sections and only joins into rounds at its top edge where it meets a pattern or a stocking stitch section. Plain borders are found in many areas and can be seen on the fisherman sitting on the lobster pot in the frontispiece.

Ribbed welts

A ribbed edge is usually joined from the very start and may be worked in a variety of ways. Most common are the simple one and one or two and two ribs, but other patterns may be found. Ribs that are wider than this tend not to be so strong, so elastic or to sit as neatly.

Decorated ribs may be found, although they are perhaps more a product of exhibition work and are therefore of a later date. Exceptions to any rule can always be found where guernseys are concerned, and the patterned welt seen on the guernsey from Eriskay in the final section was probably made between 1935 and 1950.

18 A ribbed welt

17 Plain border gansey (left) worn by George Cromarty with Tom Stevenson of Holy Island, retiring after 100 years' joint service with the lifeboat

SEAM STITCHES

Although guernseys have no seams, mock seam stitches are usually incorporated into the design and very usefully mark the sides of the garment. At best the seam stitches should flow upwards out of the welt, not simply being increased at the top of the welt as a last-minute addition.

The width of welt ribbing may determine the number of stitches that are used to form the seam. One and one rib (19) is an ideal base for a single seam stitch, which is worked as a purled stitch with stocking stitch to either side.

Two stitches can also be used, and in Caithness this seldom seems to vary. They can be worked above one and one rib, by increasing one of the seam stitches where they begin, or they can flow more easily out of two and two rib (20). It is not even unusual to find seam stitches that are decorated or

19 A single purled seam stitch

25

20 *Two purled seam stitches*

21 *Decorated seam stitches*

wider than a simple mock seam would appear to need.

Three stitches might be worked as a purl, a knit and a purl stitch making their own small column of rib, or the centre stitch can be worked alternately knit and purl on a purled background (21).

On Whalsay, to the east of the Shetland mainland, is recorded a side panel that takes the place of a seam, carrying a repeat of the main pattern and much too wide to be considered as a seam stitch replacement. In Morayshire, too, the seam stitch can be so much part of the pattern that it seems not to be extra. Here the patterns tend to be vertical panels that read round the guernsey, each separated from the next by an ornate narrow ridge or rib, of which the side ribs simply form two more, not breaking into the continuity in any way.

GUSSETS

The gusset, or gushet as it becomes in Scotland, is a means of widening the garment towards the upper chest. This gusset is then held unworked until the front and back of the body are worked separately to the shoulder. Once the shoulder is complete, stitches are picked up around the armhole, including the held gusset stitches, which are then decreased down the sleeve until the gusset stitches have all been reduced, leaving only the seam and sleeves stitches. Further shaping is given to the sleeve but now it will be worked at each side of the seam stitches until only sufficient stitches remain for the cuff, and the sleeve is the required length.

The gusset, basically a diamond shape, can be formed in many ways. The most usual is to form a

22 *Gusset outlined by a single purled stitch*

gradually widening island of stitches, starting from the centre of the seam stitches, and outlining it with seam stitches (22). This may be elongated if four or more rows are worked between shapings, or it may be squat and flat if the stitches that are increased then decreased are worked more closely together. The seam stitches may not only outline the gusset but may also be added to so that they form an outline and continue right through the centre of the gusset as well (23).

In Thurso it is usual to work the gusset differently. Here no stitches outline it, but the seam stitches pass through the centre, with the increased stitches forming a stocking stitch area or triangle to either side, before they join in a straight line to the patterned yoke (24).

This plain area of gusset is only occasionally decorated, and when not overdone, or when neatly worked, a moss stitch gusset can look attractive.

23 *Gusset with seam stitch carried through it and with the increased area also outlined by a purl stitch*

24 *Northern gusset worked on either side of seam stitches*

SHOULDERS

The simplest, and possibly the oldest method of joining the shoulders, is to place equal numbers of stitches from back and front together and cast both sides off together on the right side (25). Next in order of simplicity is a shoulder which carries the front

stitches across to the back and casts them off together with the back stitches on the wrong side of the work. This type of shoulder has an invisible seam, since the shoulder area is worked in a different pattern from the rest of the back and front, as in the modern pattern from the Outer Hebrides, where Indian corn stitch is used for the strap (26).

Another shoulder strap, worked in a similar way, is used for the Appledore jersey in the final section, where you will find working instructions and a close-up photograph of the shoulder.

An often-used shoulder pattern is the rig' and furrow, likened to a ploughed field. This east coast of Scotland version is worked in a similar way to the

26 *Contrast patterned shoulder strap from the Outer Hebrides*

25 *Emmanual Bryant of St Ives wearing a gansey with shoulders which have been cast off together on the right side*

27 *Rig' and furrow shoulder strap with grafted edge for an invisible join*

Hebridean strap, but is joined invisibly by grafting in place of casting off the edges together (27).

Pattern comes into its own on the shoulders of Caithness guernseys, where extra stitches are cast on over the top of the arm and worked up towards the neck edge, gradually working in the stitches from the side needles (28). Instructions for this can be found in the Thurso gansey in the final section.

28 *Caithness patterned shoulder strap*

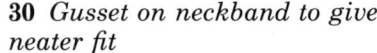

29 *Scottish buttoned neckline*

NECKLINES

Necks seem to have given a lot of trouble and were often too slack or too tight. Usually they were simply one third of the stitches of front and back, taken from the centre of the work after the shoulders had been cast off. In Scotland, a neater neckline was made by leaving the side of the neck open and working the neckband in rows, not rounds, working in buttonholes and matching these with a neat row of smoked pearl buttons (29).

Another method of drawing in the neck is to work a small gusset at the base of the neckband, gradually decreasing as the neckband progresses (30).

30 *Gusset on neckband to give neater fit*

4 Cornwall and Devon

'But I never meant to write a book. I was asked to knit a fisherman's guernsey for the Women's Institute display at the Royal Cornwall Show and I had to find out everything about them. The type of wool, where to begin, how they were made and what the patterns were like,' said Mary Wright of her own fascinatingly detailed book *Cornish Guernseys and Knit-Frocks* (31). Her 12 years of seeking, questioning, reading and experimenting, to say nothing of poring over old photographs to learn all that they could tell, must remain the key to anyone interested in Cornish knitting. It would indeed have been accidental to have found the way back into the past without her guidance and help.

Her book covers every aspect of knitting, and even to read it is to journey round the rocky coast, over sandy beaches, down deep lanes and into tiny villages where long memories can yet be stirred to remember knitting sticks and patterns, flying fingers and fine steel needles. It brings back a time before prosperity, when work went on even as neighbours met to talk, and fingers slipped loops from needle to needle as watchers kept the sea under constant surveillance for the sight of pilchard shoals (32).

For it was a very different Cornwall that existed when guernsey knitting was at its height: a time when knitting was necessary to clothe the family and could help eke out low incomes by contract work for local drapers, who in turn sold to merchants. In 1859 a draper in Liskeard sought '100 Hands to Knit Frocks', and ten years later an advertisement in Looe asked for '500 good Frock Knitters, constant employment'.

31 *Mary Wright, author of* Cornish Guernseys and Knit-frocks, *darns in ends on a finished guernsey. In the background can be seen a guernsey patterned with Cornish lattice pattern*

32 *Tucking pilchards off Cadgwith. Photo by Jordan of Truro, c.1910*

Although it is claimed that a knit-frock could be made in one week, bringing in a wage comparable with an experienced domestic servant, it was never a leisurely 40-hour week. It did, however, allow the women to be at home and undertake other domestic chores at the same time, since the knitting could be picked up at any spare moment.

POLPERRO

Local records and histories make the volume of work obvious, but tell nothing of the actual knitting or the patterns: those details that are interesting to the guernsey knitter of today. It is an incredibly slender chain of circumstances that make Polperro unique in supplying a page of guernsey history that would otherwise be blank. To walk down the steep road into Polperro is to go back in time. High rocky cliffs surround the village except on the seaward side, and the houses, if they are not actually close to the harbour, find a footing where they can, clustering close to each other as if for company. Dr Jonathan Couch, in his *History of Polperro*, for which he was making notes in 1856, wrote:

'Near the strand on the Lansallos side are the "fish scales" or market, where you may often see a busy group of fishermen, clad in Guernsey-frocks, sou'westers and sea boots, bargaining by a sort of auction, with loquacious jouters (travelling fishmongers), for the contents of the "ocean-smelling osier" (too sweet a term, perhaps, for the brown slimy pannier) or for piles of cod, ling and conger, too bulky to be so contained.'

Lewis Harding

It was Jonathan Couch who introduced a temporary patient to Polperro and it was he, Lewis Harding, who made a study of 82 fishermen of Polperro, forming one composite picture, which now hangs in the Rowett Institute (33). As a study in people, it is a work of art: as an aid to understanding more of the knitting of 1850–60, it is without parallel.

33 *Portrait of 82 Polperro fishermen, which, although faded, can still show details of many gansey patterns used around 1850 Photo by Lewis Harding*

Although at one time the name of Lewis Harding was forgotten (Andrew Lanyon's *Rooks of Trelawne* recalls the search for his identity) his photographs lived on to show that by the mid 1800s guernsey knitting made use of endless varieties of pattern. Over 100 years later it is still possible to see the stitches clearly, to reconstruct the patterns and to notice details, such as the shaping on the neck of Richard Searle's gansey, detailed instructions for which are in the final chapter.

A triangle similar to the main panel of Richard Searle's guernsey can be seen on the young street musician with the triangle (34), and next to him the boy with the concertina wears a chevron pattern which is effective in its texture (35). This chevron is to be found in many places, and, in the north of England, when worked over 13 stitches or '12 + 1' as the knitter phrased it, is said to have a religious meaning and represents Christ and the twelve disciples.

The chevron forms the centre panel on one of the very best of Lewis Harding's photographs (36). On the guernsey worn by Jim Curtis with the maid on his knee, can be seen an example of pattern combination which would be difficult to fault. Here single moss diamonds are teamed with an open diamond, unusual in its centre line (37). Standing beside him is Charles Jolliffe junior, who sports knitted braces, as well as a gansey which shows how to use rib for speed and adds only the minimum amount of pattern in the yoke border, repeating this to effect on the sleeves (38).

Contract knitters

Again it is in Polperro that another coincidence fills in a gap in the story of the contract knitters. In his writing, Jonathan Couch makes only one reference to the fact that 'the women knit large quantities of worsted stockings for the army in the Crimea'. But the returning officer for the 1851 census, John Giles, must have known much more about the knitting or have been a very exact gentleman. In his returns, 29 women and girls are noted as being knitters. Their ages are as varied as their patterns, from Emma and Sarah of Bevill Row at only 10 and 11 years old (as was Eliza Torward of the Warren), to Susanna White of Talland Street at 65 years old. Most, however,

34 *Youngsters making music in Polperro about 1850. Photo by Lewis Harding*

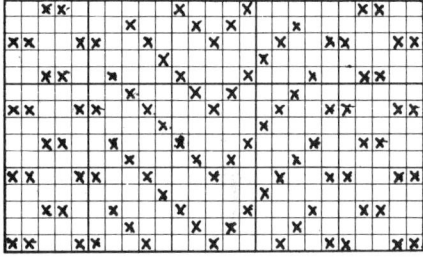

35 *Concertina player's pattern chart*

36 *Charles Joliffe and son with Jim Curtis and a Polperro maid. Photo by Lewis Harding, c.1850*

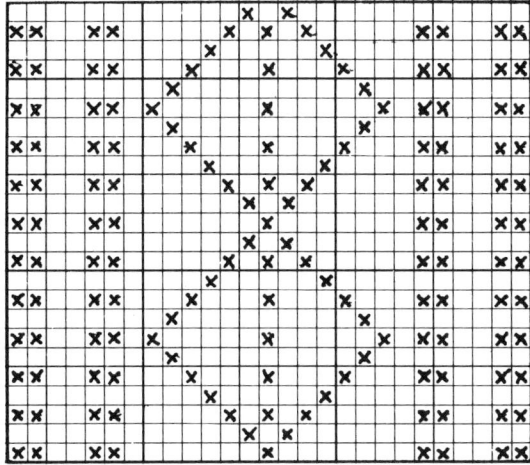

37 *Jim Curtis's diamond pattern chart*

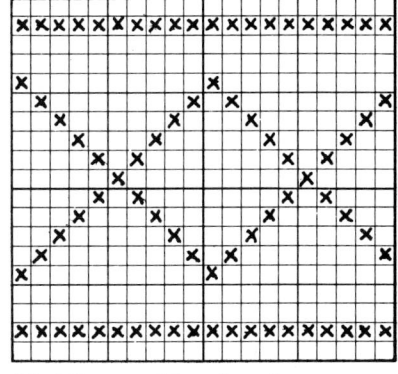

38 *Diamond border chart*

were aged between 16 and 30. Although 29 may seem a very small number when 500 were being advertised for, it is definite information, whilst the records for the other parishes around Looe and Liskeard, where there must have been many more knitters, make no mention of the fact.

PATTERNS

Leading in its documentary evidence, Polperro was by no means alone in its guernsey knitting, and Mary Wright records patterns from many places on a lengthy coastline which, at $275\frac{1}{2}$ miles, is almost exactly 10 per cent of the total for England and Wales.

Block patterns

Throughout the county, simple patterns abound, ribs of every sort and many block patterns. An enlargement of John Northcott's picture showed an interesting block pattern easily adapted to any size, and characteristic of Cornwall in its use of garter stitch ribs between blocks of stocking stitch, instead of the purled ribs that might be used today (39). Polperro was not the only place in Cornwall where block patterns were found. Old pictures of the Vicar of Morwenstow, clad in a guernsey-frock like the fishermen, because he felt that as a 'fisher of men' this was more fitting than the traditional clerical dress, show a similar pattern (40).

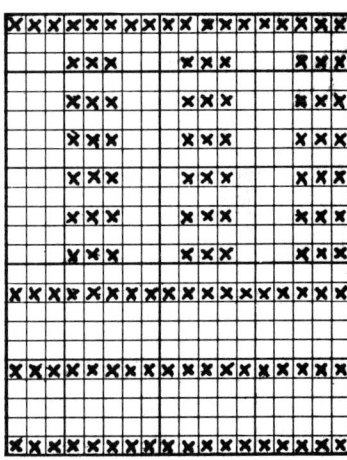

39 *John Northcott's pattern*

40 *Vicar of Morwenstow's pattern*

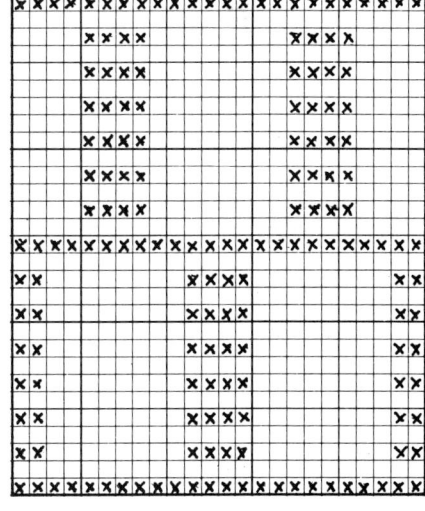

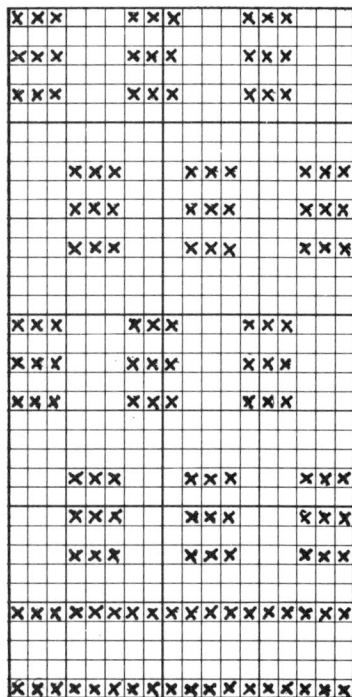

41 *Cornish lattice pattern chart*

Lattice pattern
One pattern closely related to the block patterns seems to be Cornish in origin and has the same garter stitch base, forming a lattice pattern rather than the more usual basket pattern (41). It can be seen on a guernsey in the background of the photograph of Mary Wright (31) and is one of her carefully recorded patterns. Used in the banded designs that occur in all areas, it produces a more pronounced pattern than the more often seen basket stitch, moss or rice stitch.

Rope or cabled patterns
Rope or cabled patterns seem to be fewer than in some northern areas, but Mary Wright records patterns from the Scilly Isles, from Sennen Cove and The Lizard, from St Ives and Bude and a double rope pattern from Port Isaac. The most unusual is a snake or serpent rope, twisted alternately right and left to make it look as if it lies on the surface. Worked with a ladder pattern, as it is in this version, it needs no lengthy calculations to make it usable for any size (42).

The snake is worked over a repeat of 16 sts but can be widened either on the ladder section or by working the cable over eight sts instead of six. The depth between cables can also be altered and needs to be increased if the cable is widened (43).

The movement to the left is made thus: slip next 3 sts to CN, hold at front, K next 3 sts, then K3 from CN.

The movement to the right is made thus: slip next 3 sts to CN, hold at back, K next 3 sts, then K3 from CN.

As in other areas where knitting was vital for making ends meet, children started to knit as soon as they could hold needles, and before long were able to help by knitting 'the trails', as the ribbed edgings were called (44). Nor were fishermen's sons ever too young to be clad in their own guernsey, wearing it with pride at being an equal with their father.

APPLEDORE
North over the unseen boundary into Devon, the knitting of Appledore takes on a different appearance. Here a 4-ply worsted was used to make a fine guernsey, decorated only at the shoulders.

42 *Snake cable and ladder pattern*

Instructions are given in the final chapter for an Appledore guernsey. The close-up photograph (p.106) shows the detail of the shoulder strap, worked from the front and joined by being cast off, together with the same number of stitches from the back, on the wrong side of the work. This is a marked contrast to

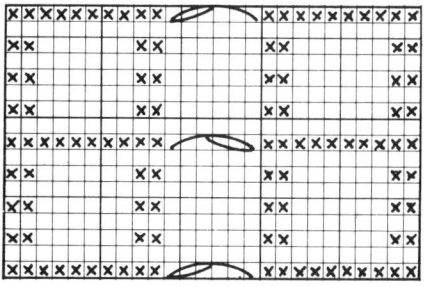

43 *Snake cable and ladder pattern chart*

44 *Children in Polperro, knitting before 10 years old. Lewis Harding c.1860*

45 *Wooden tack of knitting stick, Appledore*

the ridge of casting off used to decorate the shoulder line of the Cornish guernseys.

Knitting stick

In Appledore a knitting needle was known as a prang, and a stick as a tack. This stick (45) was pictured in Appledore and, although patterned, was unfortunately undated, unlike one of two from Cornwall dated 1797 (46). Dates on knitting sticks have their own part to play in history, and a stick found in Brixham, dated 1787, is believed to have established valuable information about the dating of the early years of the summer migration of fishermen to the Yorkshire coast. Previously the earliest known date had been around 1830, when it was known that many families, complete with furniture, would travel north for the summer months. It was while in Yorkshire that the women learned to use a stick to support their work, leaving the fingers free to move more quickly right at the needle tips.

One account of the sticks notes that 'The stick was held at the left side of the waist hooked over or tucked into a cord or sash called a cowband. One needle was held in the stick, leaving the hand free to manipulate the wool.' Abb wool stockings seem also to have been worthy of special note in Devon, where they seemed to be less common than further north. These were made from the second lowest grading of wool and were cream in colour, rich in oil to repel water. Worn inside sea boots, they can be seen clearly in the frontispiece.

46 *Two Cornish knitting sticks, one dated 1797*

5 Norfolk

Once a year, in the days of the herring fleets, Yarmouth was host to fishermen from the south, from Yorkshire and from the far north. It was also a main market for the sale of ganseys made elsewhere. Possibly because of this it is difficult to find any style that was its own. Out of Lowestoft and Yarmouth, however, in the smaller towns, in Caister and Sheringham, ganseys came into their own.

SHERINGHAM
Sheringham seems always to have been a special area for fine ganseys, and wool as fine as 3-ply has been used with needles as thin as wires (the old Nos. 16 and 15). The fine yarn made moss stitch, known as hailstones here, look not unlike icy-cold, real hailstones, for it gave a clarity which is lost in thicker yarns. Cable patterns are popular and the rope is often known as a coil of rope. Most patterns are yoked, which must make knitters wonder about a link with Cornwall. But which came first?

Brocade patterns
It is in Sheringham that the strongest link exists with possible earlier brocade knitting, for who can deny that the pattern worn by Robert 'Tarr' Bishop in the photograph of him and his wife taken in 1906, does not look like brocade (47)? The patterned yoke is edged with several stitches alternately knitted and purled for two rows, which draws in the actual depth of the armhole, allowing the sleeve to fit reasonably

47 *Robert 'Tarr' Bishop and wife, Sheringham*

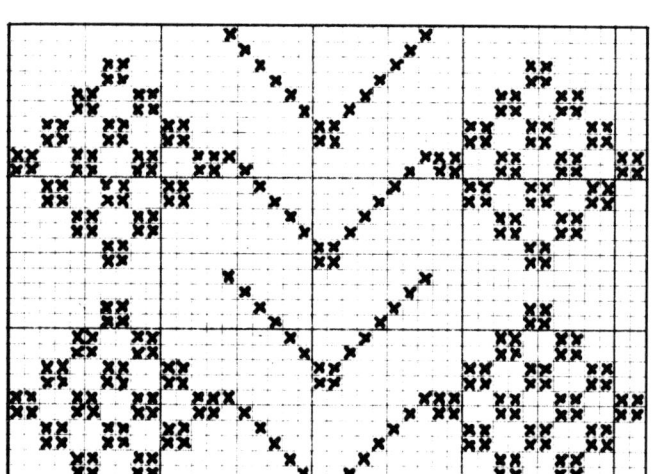

48 *'Tarr' Bishop's pattern chart*

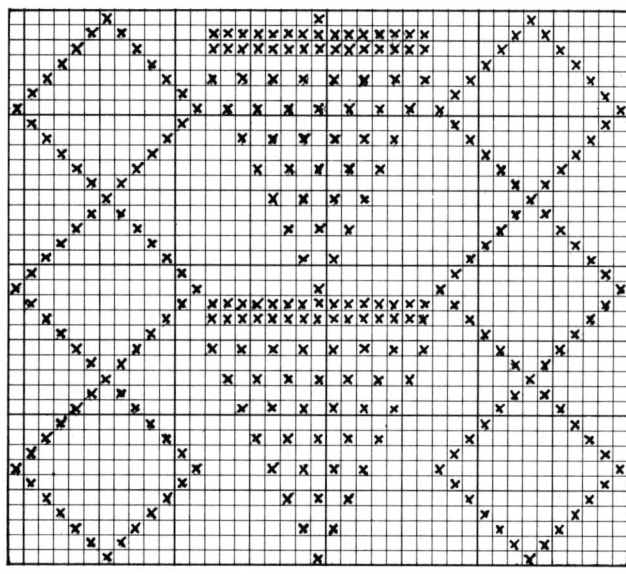

49 *Mrs Esther Nurse's pattern chart*

at the top instead of flapping too widely, as it would otherwise do (48).

Another pattern speaks of brocade and is wrongly recorded in the reprint of Gladys Thompson's book. It is a pattern from Michael Harvey's collection and was knitted by Mrs Esther Nurse of Lower Bodham in 1950 in 4-ply wool (49).

Compare this pattern with Richard Searle's pattern (158) in the final chapter, remembering that his photograph was taken almost 100 years earlier, and also with that of the triangle-playing musician in picture 34. It would hardly seem possible that they could have been knitted independently of each other, and yet there is no known link.

50 *Billy 'Clubfoot' Mayes and family, Sheringham, c.1870*

Interesting gansey patterns

A photograph of Billy 'Clubfoot' Mayes and family (50), taken about 1870, shows two interesting ganseys. At the back is a bold diamond patterned design, unusual on banded ganseys, and not unlike a pattern which much more recently has been attributed to Foula in the Shetland group of islands. Below the diamonds is a band of alternated double stitch, and above it moss stitch like the diamonds themselves. Beneath the chest band is a deeper ribbed band to the base of the yoke. On the front gansey can be seen the same ribbing but wider, possibly to accommodate the wider man, with the ribs between worked in garter stitch, not purled. Both the other bands are probably worked in moss stitch given a diagonal appearance by the light.

51 *Christopher Brown 'Cutty' Cooper, Sheringham, c.1890*

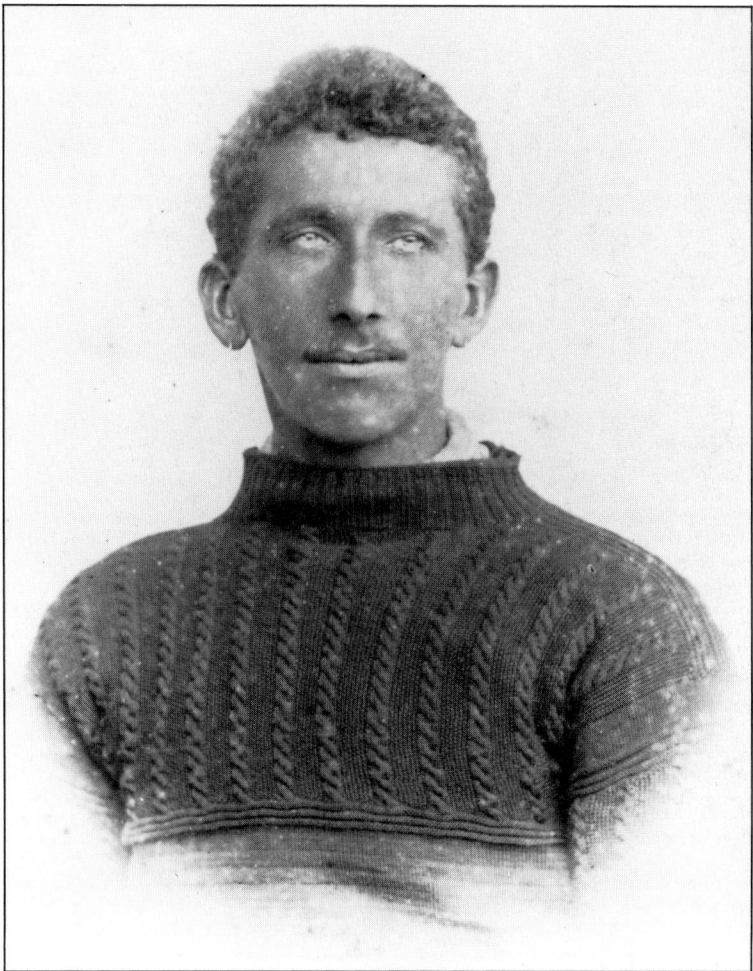

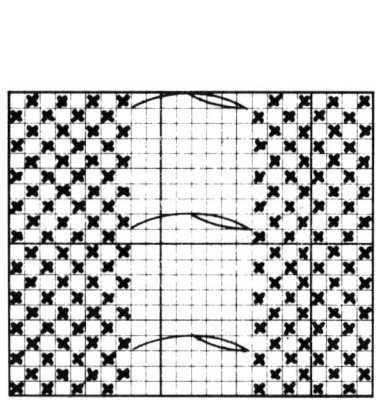

52 *'Cutty' Cooper's pattern chart*

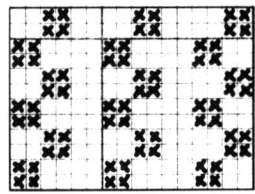

53 *Young 'Cutty's pattern chart*

It is not often that there is the chance to see a youngster in his best gansey and to be able to place alongside a photograph taken about ten years later, when he is again wearing his best gansey. The picture of Christopher Brown 'Cutty' Cooper (51) was taken around the early 1890s. The gansey he is wearing was probably knitted by his mother, and it is a perfect example of hailstones and coil of rope (52). An earlier faded photograph of 'Cutty' with his parents showed an unusual small pattern which cropped up again, not without interest at Flamborough Head (53).

Many times has Sheringham promenade acted as a photographer's studio, but seldom to better purpose. Taken in 1906, this photograph shows all three ganseys to great advantage (54). If only the other two fishermen could be asked to take off their 'slops'! The man on the left and the standing man in the centre are both wearing patterns that are related to the pattern in picture 47. Full working instructions are given in the final chapter for the standing man's gansey, while the other is shown on the chart (55). Second from the left is a pattern similar to 'Cutty' Cooper's pattern in picture 52.

Again on the promenade but during the 1920s, with papers to prove it, as well as the beach inspector and a lady visitor for interest, are two patterns, one of which is typical and one of which is much more unusual. The two fishermen at the right are Jimmy 'Coalie' Cooper at the end and next to him James Dumble (56). James Dumble's pattern is unusual in its arrangement and perhaps misleading to chart, possibly because the yarn tension is not quite as usual (57). 'Coalie' Cooper's pattern uses the Sheringham hailstones with a chevron on either side, ribbed with purl (58).

In both of the last two pictures there are other ganseys, but in each case these appear to be

54 *Sheringham fishermen, 1906*

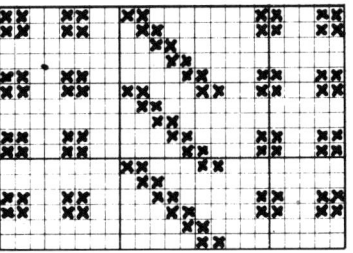

55 *Line and block pattern chart*

56 *On the promenade, Sheringham, c.1920*

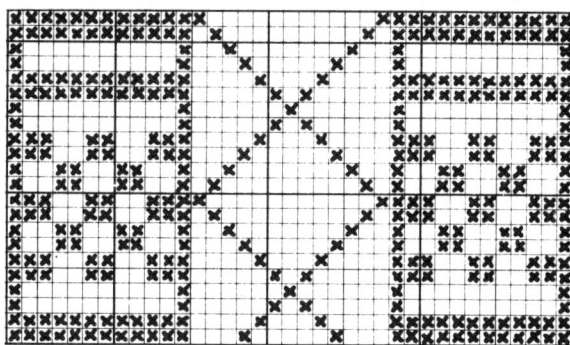

57 *Jim Dumble's pattern chart*

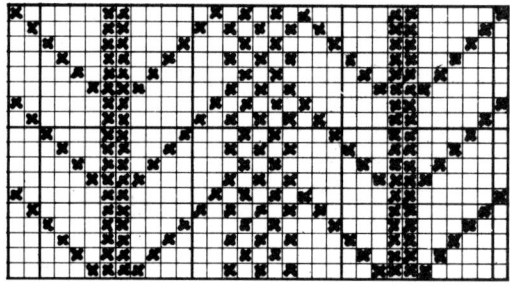

58 *Jimmy 'Coalie' Cooper's pattern chart*

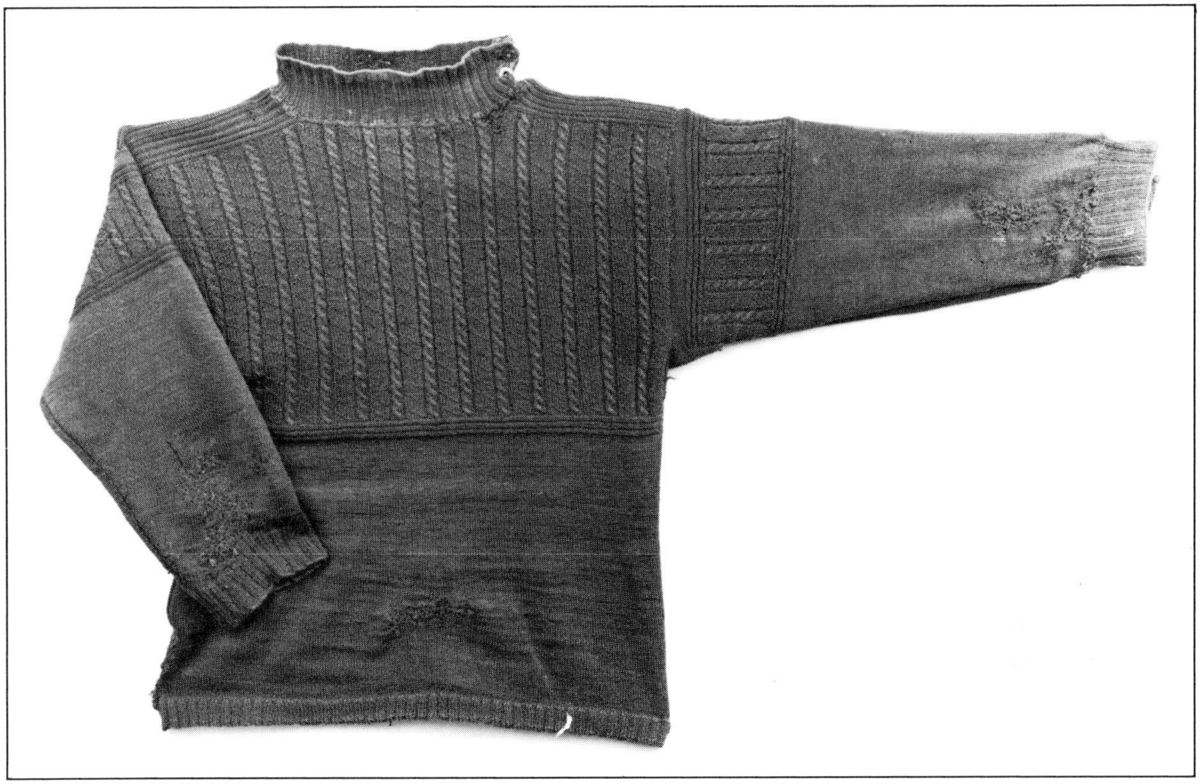

machine-knits. They can often be difficult to detect, some of the patterns being very like hand-knitted patterns.

Coils of rope were important in the Sheringham ganseys, and although they varied in width they twisted in one direction only. Henry Valentine Little, known as 'Pinny' Little, born in 1892, lived for 73 years and was both fisherman and lifeboat man (59). A gansey which belonged to him shows a fine rope four stitches wide, twisting to the left, and panels with single purl stitches (60). Unusually for this area, it has neck buttons, which may well be Scottish influence.

CAISTER

In marked contrast is a pattern from Caister which, although the rope is worked over only six stitches, has a decidedly chunky appearance (61). This is due to two factors. The rope is twisted every 16 rows, allowing the stocking stitch area depth in which to reach its full width before being drawn in again at the next twist. The fabric at the sides of the rope is

59 *Henry Valentine 'Pinny' Little's gansey*

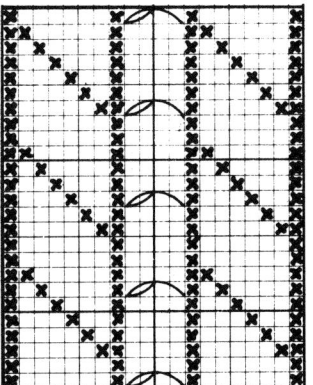

60 *Henry 'Pinny' Little's pattern chart*

47

61 *John 'Snouts' Cox's gansey, from Cromer, before 1938 when he died*

63 *Alex Kerr, lion-tamer, in a gansey knitted by his Sheringham born mother*

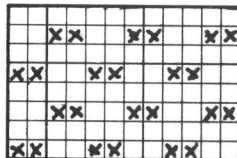

62 *Background pattern chart for 'Snouts' Cox's gansey*

worked in garter stitch, which draws the rib down, making it less in length than the stocking stitch section and so accentuating the puckered effect.

This gansey belonged to John 'Snouts' Cox, who died in 1938, and shows clearly one of the simple patterns closely related to moss stitch (62). Sometimes worked with single purl stitches alternating on every second row, this version alternates two purled stitches, which stand out very clearly in the fine yarn against their stocking stitch surrounds.

In the world of guernsey knitting, it is unwise ever to claim that any detail or pattern proves that it was made in any one area. It has been written that all English cables are edged with purl stitches, but Scottish cables may have stocking stitch or moss stitch worked up to their edges. This leaves Norfolk and Cornwall in some other land altogether. It is also claimed that only fishermen and their fishing sons wore guernseys, but this boldly yoked pattern is proof that this is not so (63). Knitted by Mrs Kerr, who was brought up in Sheringham, it is being worn

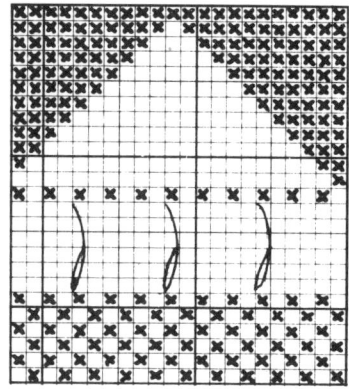

64 Mrs Kerr's pattern chart

65 Cromer Lifeboat Crew Dinner, held at the Metropole Hotel in Cromer, 1928

by her son Alex, a lion-tamer by trade. Mrs Kerr also spent many years in Scotland and this guernsey shows a perfect marriage between the yoked shape of Sheringham and the wide flag pattern found as far north as Caithness (64). Where one piece of evidence can cause confusion unless all its surrounding facts are known, there are always those that serve to bolster other beliefs.

In nearly all communities, guernseys were worn for work, with the newest one kept for best until the next in line took its place. In many areas it was also worn for the wedding ceremony – the bride wearing a special dress, which would then be worn for best for many years to come. In the areas researched there is no proof, as has been claimed, that fishermen were even buried in them. That they were worn for best has been shown again and again in carefully posed studio photographs taken with the family or while away and sent home. Cromer can confirm that it was indeed worn for best at no less an event than the

Lifeboat Crew Dinner held in 1928 at the Metropole Hotel (65).

The rapid increase in the tourist trade brought many families to the area, as early as 1890, in search of fresh air and seaside pleasures, such as an outing on the *Agnes Etta*, weather permitting. Despite the tourism and the income it brought, however, times could be hard and the fisherman remained thrifty. Even knitting needles were a needless expense if an old, steel ribbed umbrella could be found. Once in the cupboard it was easily at hand if a needle became bent or a new set was needed and it took only minutes to point up the tip and have them ready for work.

6 Yorkshire and Northern England

Yorkshire begins on the north bank of the Humber estuary, or did until the boundary changes, and for the majority of people perhaps it still does. For gansey history the river is important for here the keelboat men wore ganseys. The last knitter of these ganseys seems to have been unique, for she must have been one of the few gansey knitters to live out of sight or sound of the sea. Mrs Phoebe Carr lived at Thorne, near Doncaster, and was much esteemed for her knitting. The pattern was similar to the yoked ganseys of Norfolk and often carried ropes and diamonds, but she also used a most unusual star motif in panels, worked in double moss stitch (66).

From Spurn Point, right up the Yorkshire coast and on into Northumbria, every fishing village has known, or still does know a great deal about the knitting of ganseys. Travelling north, Flamborough Head is not just a geographical landmark but is also a landmark in the story of the guernsey, for this marks the point where stitch patterns and styles are very different from those in the south.

FLAMBOROUGH HEAD

Donkeys no longer tread the steep path from beach to cliff top with their fish-laden panniers, but your first experience of the North Landing may well include a glimpse of a gansey going about its duties by the lifeboat, which stands bright and ready in the morning sun, dwarfed by high chalk cliffs that cut off all but the open sea. One of the main Flamborough patterns must make it one of the most outstanding compositions in textures of all time (67).

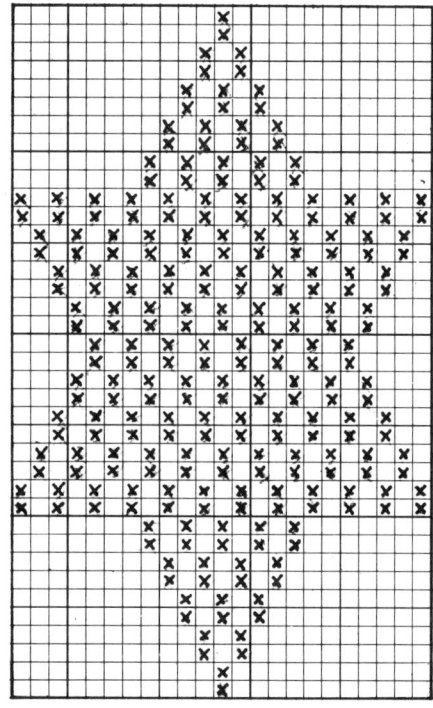

66 *The Humber star pattern chart*

Mr Shepley, who lives on the cliff top, leant against a boat to show off the contrast between birds' e'en (moss stitch), rough with its pin-head precision, and the smoothness of stocking stitch, the twist of ropes and the lattice-like patterns of open and double moss diamonds. A centre panel carries a double moss diamond, and on either side of this is a wide moss stitch panel, stocking stitch edged and with a left twisting rope in its centre (68). Beyond this again is a panel with an open diamond. Depending on the width required, each panel can be altered to make the necessary adaptations.

The older photographs recall a time when patterns were simpler, and the gansey worn by John Knaggs (69) as he works at his pots, shows birds' e'en and a narrow rib with stocking stitch edge and one repeat of a basket stitch pattern in the centre (70). The sleeves are patterned half-way from the top with a simpler variation of the main pattern. This pattern caused a great deal of interest a few weeks later when it was realised that this pattern resembled the yoke pattern on 'Cutty' Cooper's gansey in Norfolk (51).

It was Mrs Scales, daughter of John Knaggs, who provided the clue, although she was quite unaware of it. She was recalling the busy days when her father

67 *Mr Shepley of North Landing, Flamborough, wearing the traditional Flamborough pattern in 1983, recorded by Gladys Thompson over 30 years ago and in use before that*

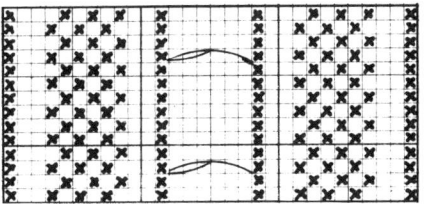

68 *The Flamborough moss panel chart*

69 *John Knaggs of Flamborough*

had been fishing, how they had all helped with so many chores, how clothes were made to last and mended and mended again. When bait was mentioned, it seemed like another language and explanations were needed to discover that 'flithers' were limpets and 'queenies' were scallops. And it was the bait that linked the two patterns – John Knaggs' and 'Cutty' Coopers – because bait for Flamborough came from Sheringham and from Wells. It seems, therefore, that knitting patterns, too, were shared between the two communities. The shoulder of John Knaggs' gansey is worked in what is called rig' and furrow, like the ridges of a ploughed field (71). These are worked by alternating two rows purl and two rows plain with the knit side facing. This type of shoulder is often cast off on the wrong side so that the join from front to back is hard to distinguish. Another simple pattern from the same area combines the moss or bird's e'en pattern with a narrow ladder-like panel.

Cliffs form the background to a beach picture from the North Landing and show a rope pattern panelled most effectively with 'Betty Martin' (72). This easy-to-work stitch is used in many parts of the north and in Scotland, although no one seems to remember who Betty Martin was or even where she came from (73). The cable is worked over six stitches and twists to the right.

70 *John Knaggs' pattern chart*

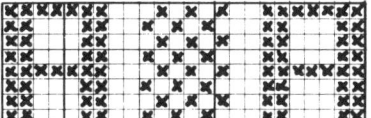

71 *Old Flamborough pattern chart*

72 *Albert Duke at North Landing, Flamborough wearing a pattern of ropes and 'Betty Martin'*

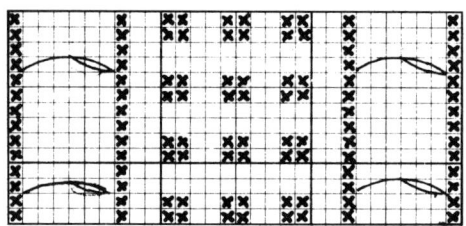

73 *Rope and 'Betty Martin' chart*

74 *George Mainprize, born in 1875, wearing a Flamborough pattern*

75 *George Mainprize's pattern chart*

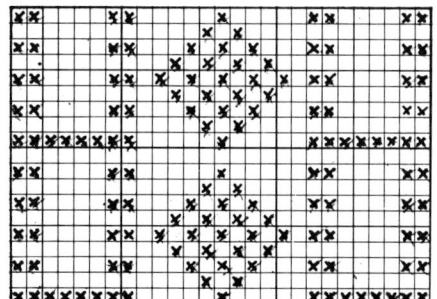

In Bridlington, Mr Mainprize went back many years in talking about his father, George Mainprize, born in 1875, who had fished from Flamborough. As a small boy it had been his greatest ambition to get his first gansey like his dad's, not so much for the gansey but to wear the welt turned up as his father did (74). He reckoned he was only seven or eight years old when he first had one and it was not knitted locally but came from Scotland and was the introduction of more varied and larger patterns. His father's pattern uses a small moss stitch diamond and ladder stitch in alternated panels and gives a surface texture that is rich and easy to follow while being knitted (75). The picture also shows the stocking stitch section knitted above the ribbed welt, before the start of the pattern, which allowed easy renewal of the ribbing when it became worn, without giving the difficult job of picking up stitches from a patterned round. Not only were the stitches easier to pick up, but a thread could be pulled out without being entangled in pattern, removing the worn section in one piece.

    Mrs Mainprize was also a knitter and had two interesting knitting sheaths, one a sheath and unusual in shape (76) the other a stick dated 1787 (77). Sticks of this sort were called 'fish' in Cornwall, but it was in Hull Museum that a 'fish' like a fish was found (78). Perhaps it was just such a fishy 'fish' that S.W. Paynter was describing when she wrote of the dame school in her *History of St Ives*: 'I must not forget to mention that tied round her waist was a wooden fish, in the open mouth of which the end of one of her busy knitting needles rested'.

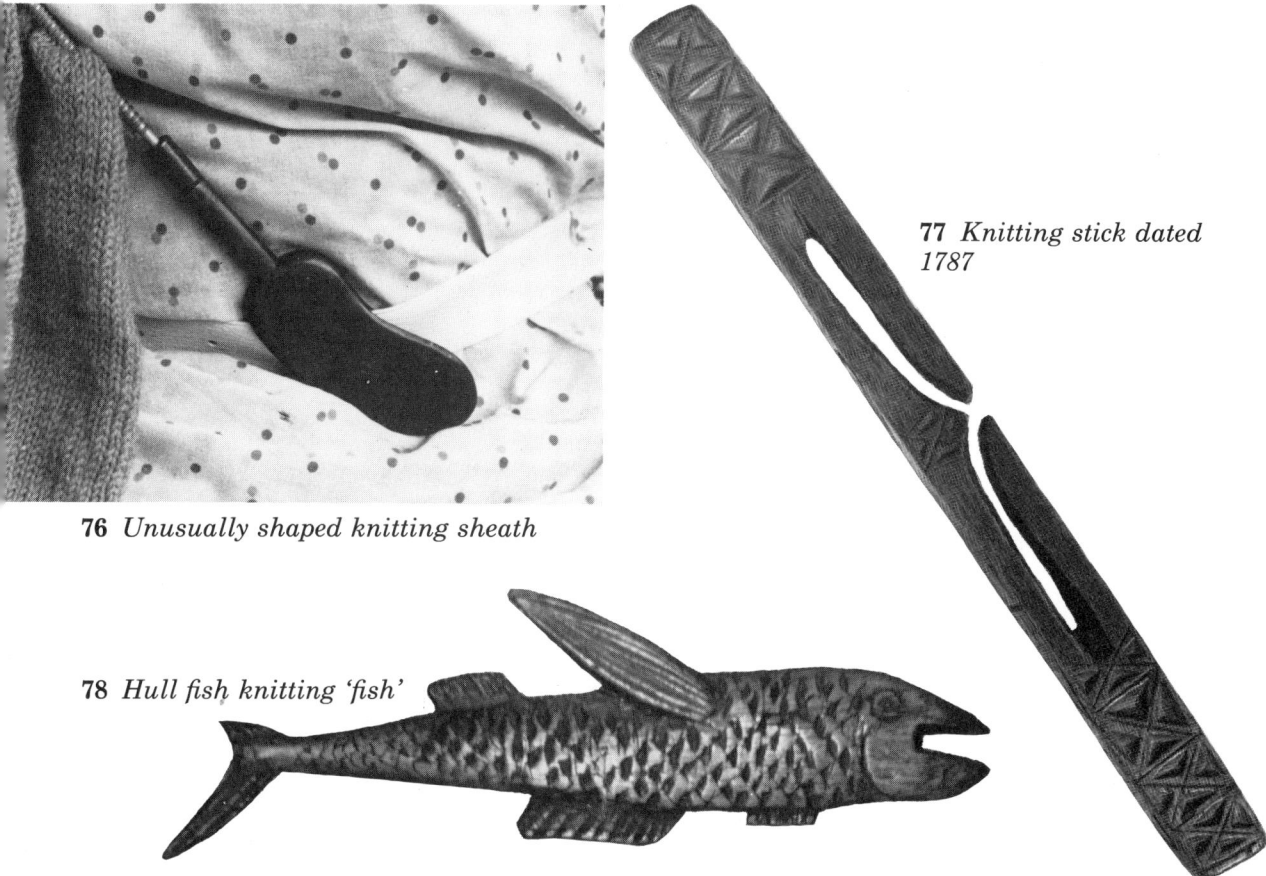

76 *Unusually shaped knitting sheath*

77 *Knitting stick dated 1787*

78 *Hull fish knitting 'fish'*

79 *Traditional Flamborough pattern showing initials knitted into the plain section above the welt*

80 *Flamborough rope, basket stitch and open diamond pattern chart*

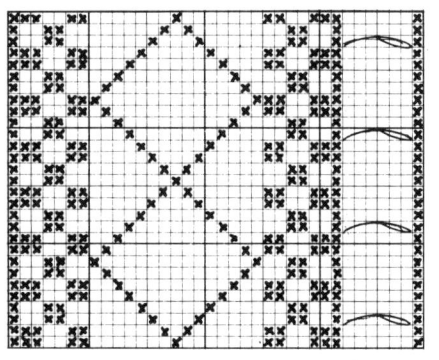

Ask an old gansey knitter how to begin a gansey and you will be told to cast on (or make) 15 score loops for the rib and add another score for the body. It had been said so often but it was in Bridlington that it was pointed out that it was easier to count by the score. In older times hooks on lines were counted by the score, bait was counted off for each hook by the score and so loops too were reckoned by the score from habit.

Carol Walkington and her husband Fred, coxswain of Bridlington lifeboat, carry on the old traditions and provided many of the details of the Bridlington

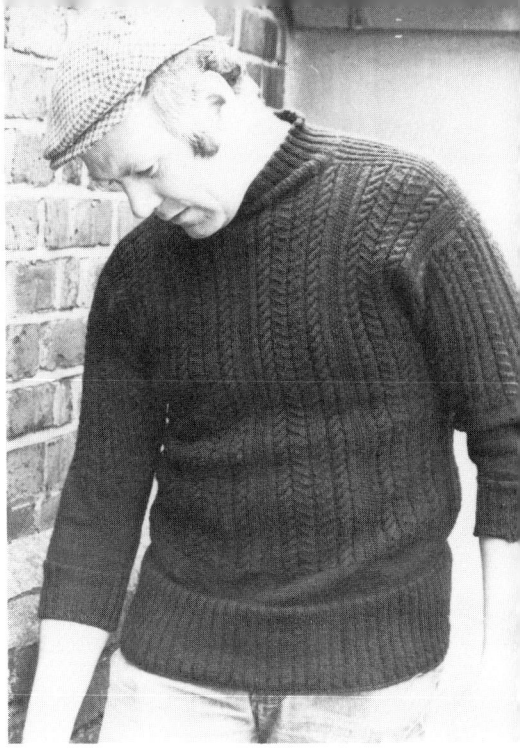

and Flamborough area, like the example of the welt reinforced by casting on with two strands of yarn which are then used for the first six or eight rounds, before continuing with single yarn (79). Carol also reverted to the old method of adding the wearer's initials on the stocking stitch section above the ribbing. The rope on this and the following pattern are worked over six stitches and twist to the right. The chart shows the pattern on the light version (80) but a similar pattern on the dark version replaces the four alternated stitches with single moss stitch (81). 'It's not a gansey unless it has ropes' was Fred's own comment (82) and the pattern of the chevron and Flamborough moss panel could prove his point (83). Carol also used the old heart pattern, found further

81 *Flamborough rope, moss stitch and open-diamond pattern*

82 *A pattern from Flamborough that 'makes a real gansey' for its wearer, Fred Walkington, coxswain of Bridlington lifeboat*

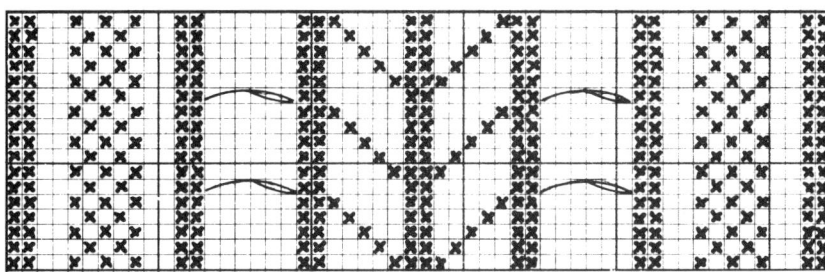

83 *Chart for chevron, rope and moss pattern*

84 *Chart for alphabet*

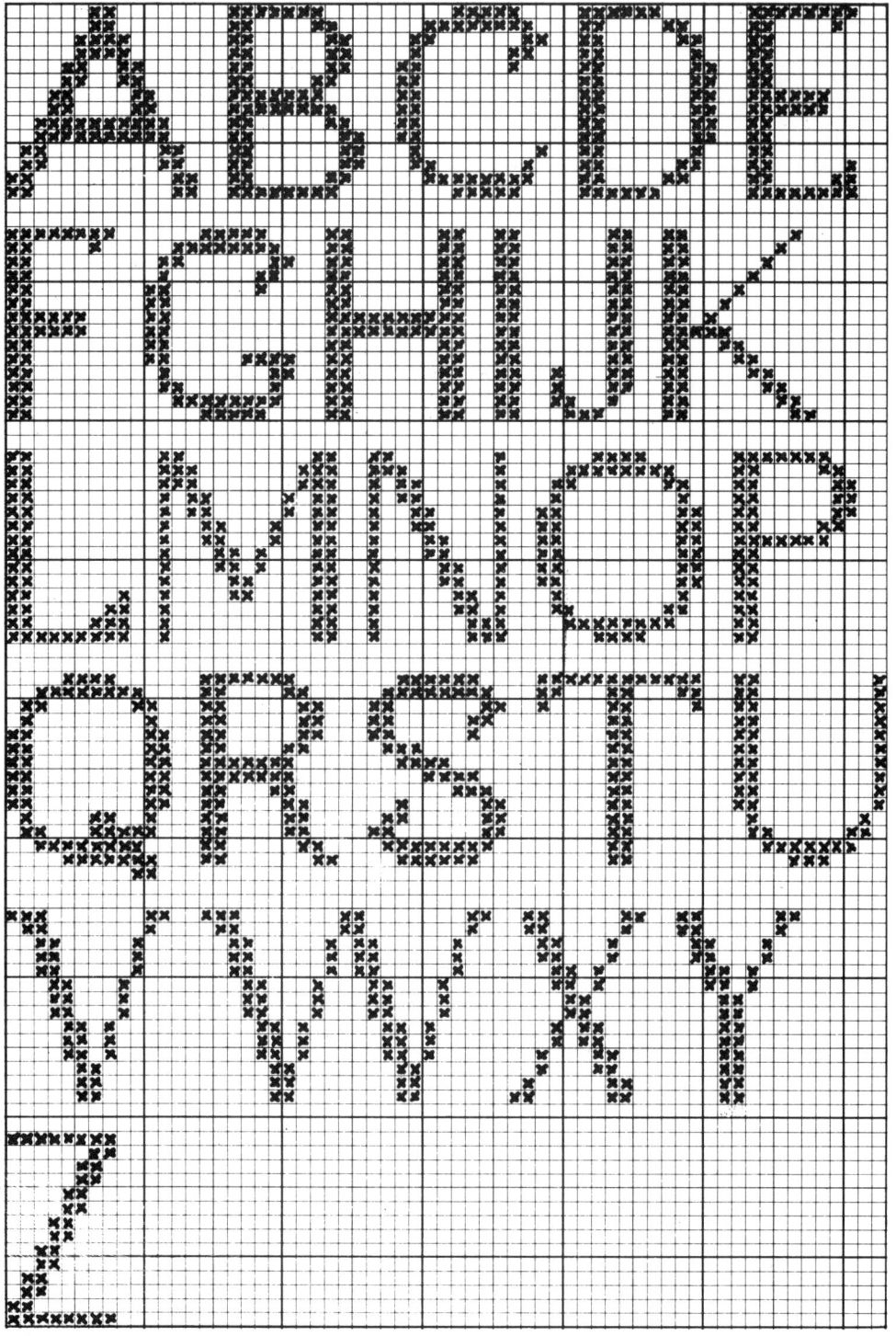

north as well, to make ganseys for all the family, and a similar pattern is given in children's sizes in the final chapter.

Give your gansey that specially individual touch of having knitted initials above the welt. A chart of letters should help you or you can use this to make up your own, possibly in moss stitch rather than just in purled stitches (84).

FILEY

Unusual circumstances often provide a means of gathering together more information than could have been hoped for, whether it is meeting a far-seeing historian, a knitter who seeks on and on to learn more, an exceptional photographer or someone who likes to keep the records of a town in order. Filey was no exception, and without the help of Mr Crimlisk, whose family came from Ireland at the beginning of the last century, so many questions would have been left unanswered. Births and deaths, names and photographs, all were at his fingertips: not impersonally, as a well-programmed computer or good filing clerk might be able to present them, but with the certainty of understanding every item that had been recorded.

In *Guernsey and Jersey Patterns* Gladys Thompson had mentioned Lizzie-Ann Pashby, but it was Mr Crimlisk who provided the contact with her step-grandson by marriage. But for Lizzie-Ann, many knitters would be less able to make a gansey for it was her patience that taught Gladys Thompson many of the patterns that have been published. Mr Willis, now a widower, spoke of her as his 'wife's step-granny', and drew a picture of a strong Yorkshire woman, endlessly energetic and capable and with time to help out with cleaning, white-washing, visitors, gansey knitting, and to have patience left over to pass on the stitches she knew so well. The passing on must at times have provided problems, and a relation or neighbour present would step in with a translation, for her rapid colloquial Yorkshire dialect could become ever broader when she felt so inclined. That she was able to carry out any of these tasks, let alone knit ganseys, was all the more remarkable for she had only one hand, but it did not affect the speed with which she could knit, and gansey after gansey came from her busy needles.

85 *Two Filey fishermen wearing simple gansey patterns*

86 *Garter stitch and rib pattern chart*

Eleven years a widow, Elizabeth Ann Pashby died at the age of 72 almost half a century ago, in the summer of 1935. On her grave stone are the words 'God's finger touched her and she slept', but during those years how many knitters have had pleasure from the stitches she handed on and, even without knowing who she was, were glad that someone with her skill had lived.

Filey is far enough from Flamborough Head to show new patterns and near enough for them to be related in type. If there is a border line between the use of knitting sheaths, sticks and pads perhaps it is here, although with the movement of people from one area to another all methods may be found

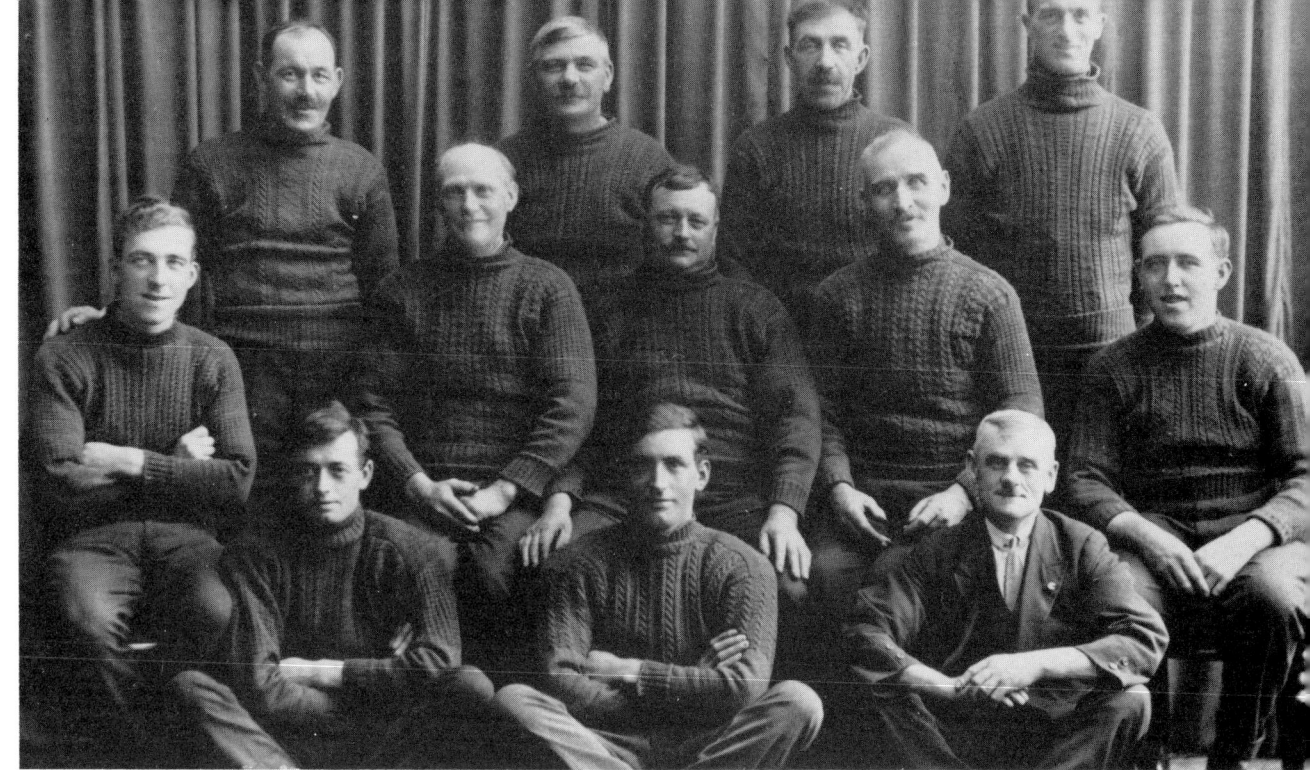

87 *Filey fishermen's choir*

throughout the country. Mrs Scales spoke of her mother using a 'tippee' or horsehair-filled leather pad, pierced with holes to hold the needle end. The pad is strapped round the waist and it is widely used throughout Scotland, where it is known as a 'wisker'.

From 1840 the yarn used was invariably Poppleton's 4- or 5-ply guernsey, known commercially then, as now, as '0½ worsted'. The yarn before that had been called simply 'wassit' or worsted, in some places as 'seamen's iron', and even the Channel Islands guernsey was made in Yorkshire wool.

Two elderly fishermen by the lifeboat station show the simpler lines of the vertical patterns often used in Filey (85). Moss stitch or garter stitch panels seem to be hardly patterns at all and yet, arranged in panels like both of the ganseys, they can be most effective (86).

With a fishermen's choir on your doorstep as a continual reminder of untried patterns or future possible adaptations, how can the Filey knitter ever fail to be inspired beyond those of other villages (87)? Even the most ardent gansey knitter must find this provides innumerable patterns.

The man at the back left is wearing a Flamborough pattern, but the others are wearing the smaller vertical patterns of Filey and most have half sleeves worked in 'Betty Martin' pattern.

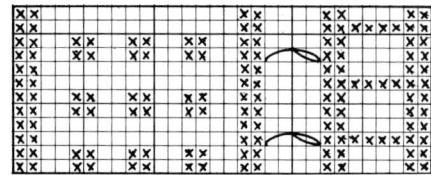

88 Filey rope and step pattern with seeded panels

89 Filey rope, 'Betty Martin' and ladder pattern chart

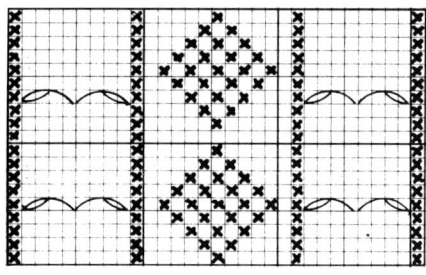

90 Filey print o' the hoof and diamond pattern chart

The fisherman at second left on the back row shows a small rope and step pattern, but the background shows a new variation of moss where the 'seed' stitches are in line, not alternating as in true moss (88). The man at the right end of the back row shows a particularly interesting design where there are steps and ladders with a panel of alternating double and single diamonds (89). In this case the same pattern is used on the half sleeve.

In the centre row are three ganseys all with moss stitch diamonds, but the second from the right also has print o' the hoof rope pattern which, although found south and north, is only used occasionally compared with simple rope patterns (90). It is in fact made by placing two cables together. Here it is worked over a total of eight stitches. The first four are twisted left by placing two stitches on a cable needle, holding them at the front, knitting the next two stitches and then the two from the spare needle. This is then reversed for the second four stitches by holding the first two stitches behind and knitting the next two stitches, then the two stitches from the cable needle.

The diamond and rope, without the added cable, twist the six plain stitches to the left (91).

Another new pattern appears in the pattern of double zigzag, rope and seed stitch panels (92). Slanting lines are often used in this way, singly, in pairs or with stitches separating the lines.

It is unfortunate that one of the best photographs of a Filey gansey shows more of the life-jacket and its cork blocks than it does of the pattern (93). The

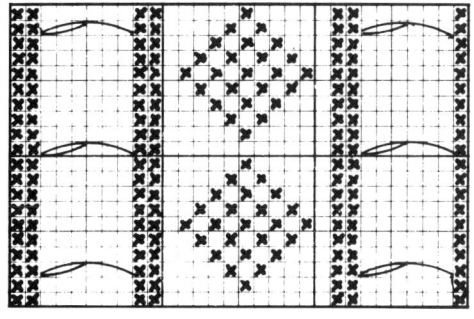

91 Filey diamond and rope pattern chart

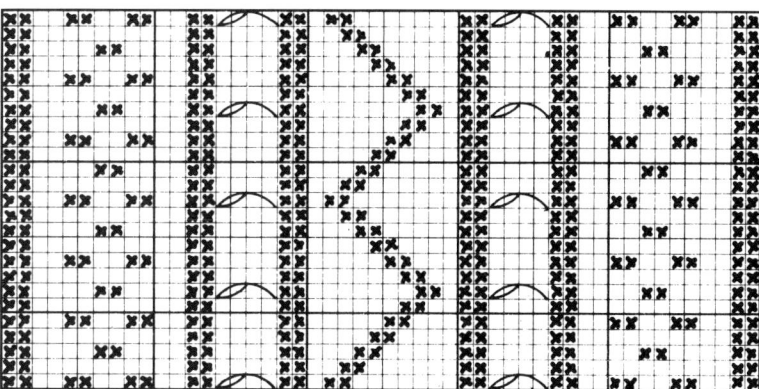

92 Filey zigzag, rope and seed pattern chart

93 Filey lifeboat man wearing a gansey with moss and purl diamonds which shows the clarity of the stitches in the wool of those days

94 *Chart for alternating moss and purl diamond pattern*

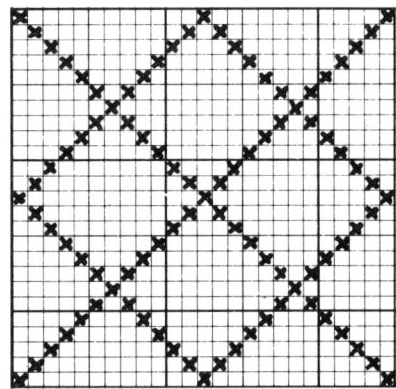

95 *Chart for diamond allover pattern*

97 *From Robin Hood's Bay another pattern of fern panelling shows how a pattern can be used in different variations without losing its character*

alternating moss and purl diamonds just show with the panels of double moss stitch between and the half sleeve which is worked in rib (94). This ribbed sleeve top allows width without bulk and makes an excellent garment, the clarity of which cannot be matched with the wool of today.

ROBIN HOOD'S BAY

To arrive in Robin Hood's Bay by car is to descend with unexpected speed from a feeling of travelling along the top of the world to being face to face with the cold North Sea. If it is at all stormy then the face-to-face aspect can be very close, with breaking waves at the street end certainly face high. The houses in Robin Hood's Bay are like mussels on

96 *Diamond and fern panels on a gansey from Robin Hood's Bay*

67

rugged rocks from which they will not be moved. With no room to move sideways, the inhabitants have found every way of making room for themselves and the houses give the appearance of creeping up on each other.

This same closeness and inter-relation is reflected in the ganseys. Most knitters will have experienced how, while knitting one garment, an idea grows from the pattern they are working, that will become their next piece of work. It is not often that a village provides a perfect example of this without having to seek it out, but in Robin Hood's Bay old photographs show an interesting line of thought. Possibly beginning with the allover diamond pattern (95), it becomes the diamond and fern (96) worn by an elderly fisherman, so like the fern in the picture of Richard Searle from Polperro in the final chapter (158). The knitting may not all have been worked by the same person, but the fern on its own (97) with the interesting decoration at its base seems to be at least closely related (98).

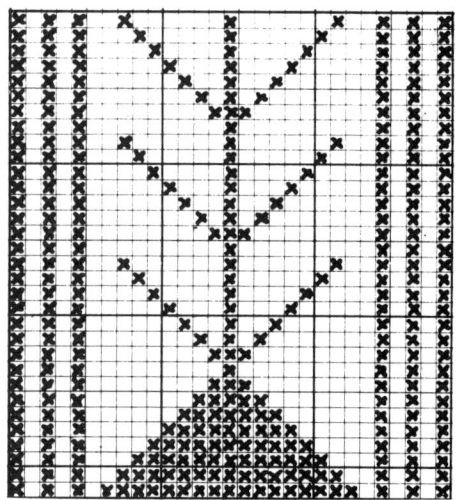

98 *Fern panel chart*

7 South and East Scotland

To cross the border from England into Scotland is to find yet another change in style, in patterns and even in name, for here a guernsey may be called a guernsey, a jersey, a gensey or a gaansey, depending which area it comes from, although all are made in the same way.

Many of the patterns used in the south can be found again throughout the length and breadth of Scotland, if in slightly different form, but there are different motifs and patterns to introduce as well. In the south the designs are perhaps more simple than in the north and western isles, where they are often exceedingly patterned and well balanced, as is shown on the front cover design.

There is also one other difference in Scotland, from the point of this book, and that is the lack of photographs other than those from private sources. Photographs taken after the early days of glass plates make dark wool very black and dense, losing all the pattern definition. There were several good photographers, one of whom, the equal of Lewis Harding in Polperro, preferred to concentrate his efforts on portraits of the upper classes (no doubt infinitely better paid), while a second, although making a fine record of commercial fishing, excelled in placing boats and people at a distance, and so again losing detail. There is one large bequest of glass plates, which must hold a great many photographs that would extend our knowledge immensely. This has been donated to a museum for this very reason, but is being withheld by its trustees from the researcher until they are catalogued.

99 *Mrs Laidlaw's pattern chart*

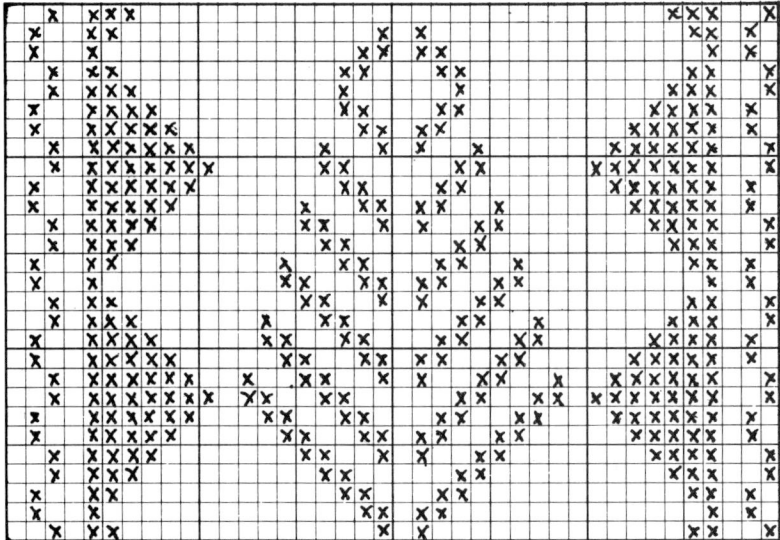

But the number of patterns available from knitters, past and present, is very large and shows that individuality has counted for many years. There has been great vying between knitters to produce, within the tradition of the guernsey, patterns with greater texture and balance and subtle use of the motif shapes. This has been helped by the fact that even now, in most years, there is a class for guernsey knitting at the Royal Highland Show. Sometimes the designs are so successful that they remain classics, like the pattern knitted by Mrs Laidlaw who came from Eyemouth, although she lived in Seahouses, and whose pattern was shown in London in 1938 at the National Federation of Women's Institutes exhibition (99).

This pattern was placed in panels round the guernsey, just like the garment worn by Skipper John Reas of Cellardyke in Fife, pictured in his wheelhouse aboard a steam drifter just prior to 1914 (100), and is typical of the use of pattern in south and east covering all the front and back from the top of the ribbing, often repeated down the top half of the sleeve (101).

EYEMOUTH

Eyemouth is the first harbour in the Firth of Forth estuary, just south of St Abb's Head, and is a joy to stop in today, when harbours may be busy with weekend craft or tourists but sadly empty of the

100 *Skipper John Reas of Cellardyke*

101 *John Reas' pattern chart*

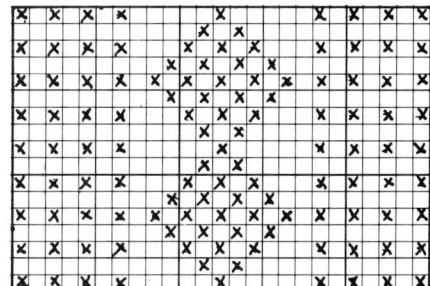

71

102 *Fishermen in Eyemouth Harbour*

bustle of landing catches or the air of expectancy as boats prepare to sail. Its quays are still busy, there are fishing boats in the harbour, if fewer, bright with plastic floats and new paint and screaming gulls circle overhead, forever scavenging.

Taken in Eyemouth harbour, this group of fishermen also shows the difference of allover pattern, with the guernsey on the right even having patterned sleeves. The pattern on the front left is worked in the same way as the Caister pattern (61) although there are many more ropes across the front (102).

To stand in Eyemouth, the harbour behind you, looking out over a calm Firth, hardly broken by waves on a sunny autumn morning, is peaceful and full of well-being. It is the sort of morning that makes the housewife tackle extra chores with ease, and that makes people feel glad to be alive. It was just such a morning in 1881 when the boats left harbour, but before they could reach the shelter of their harbour, the picture had changed. Furious waves picked up the boats as they returned, crashing them against the rocks, dragging away the men who could not even reach the arms held out to rescue them. That storm took the lives of 129 Eyemouth men – one half of the male fishing population – and was responsible for a total of 189 men being drowned.

In the days of herring fishing the fish had to be followed, and men and women gutters would travel to wherever there was work. 'Schools of herring lodge within the Firth of Forth from January until March, while they are to be found in the vicinity of the north-eastern coast of England from April until June; then from July until September they visit the waters of the German Ocean in the vicinity of Eyemouth. The months of October and December see shoals of herring in the south, near to Yarmouth and Lowestoft.'

The north shore of the Forth is also lined with small harbours that have existed for hundreds of years. There is Buckhaven and Pittenweem, St Monance and Crail, Anstruther and, round the east neuk of Fife, there is St Andrews.

FISHERROW

In Fisherrow, in the ancient burgh of Musselburgh not many miles from Edinburgh, the first Friday in September sees the fishermen and women celebrate the end of the season for another year. As long ago as the fourteenth century there was fishing at Fisherrow, and the fishermen are proud of their heritage, even if the harbour is quiet and the boats fish from elsewhere.

The Fishermen's Walk is an old tradition, broken only during the war years, when the fishermen in

103 *The Walk, from Fisherrow through Musselburgh, held originally to mark the end of the herring season, when jersey clad men lead the brightly dressed women along the main street as far as the eye can see*

their dark jerseys march through the town to Pinkie House, where they hold their annual sports (103). It is a fine sight to see the men bearing the banner presented in 1868 by Lady Hope and bearing the words beneath Fisherrow Friendly Society 'Weel May the Boat Row'. Their history is also remembered in the medallion worn by the Walk Leader, for it was presented to them for their offer of help during the time of the Napoleonic Wars, and bears the date 1796.

Behind the jersey-clad men come the fisherwomen in their full striped skirts and bright blouses topped by patterned shawls, many of which are family heirlooms, handed down from mother to daughter. A flurry of white handkerchiefs wave to friends and neighbours as they wend their way from Fisherrow, through the main street of Musselburgh and past the old tolbooth with its tower, said to precede the invasion of 1544, and which still houses the clock presented by the Dutch in 1496, when there was much trading between the two countries. A sharp right turn and the smiling, happy throng enter the

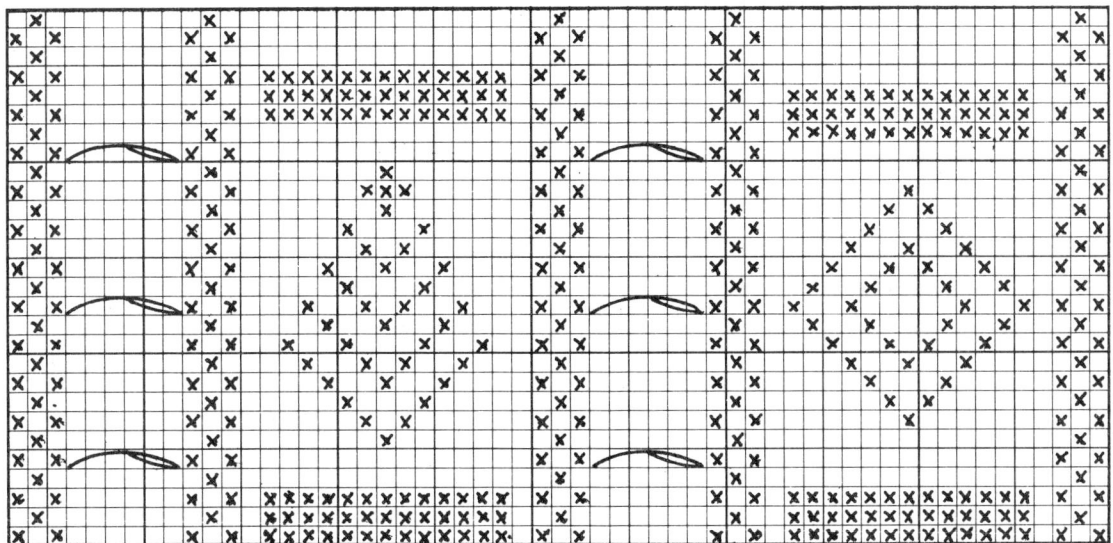

grounds of Pinkie House for an afternoon of entertainment before a well-earned tea fit for the occasion.

It is also an occasion that seems especially blessed with dry weather. It may rain lightly in the early morning, to lay the dust as it were, but it never rains from lunchtime onwards. One old fisherman, asked by Lord Elphinstone why this was so, had the quick wit to say that that was handed down too.

Here the panelled jersey can often be additionally textured by having diamonds worked, each in its own box, the panels separated by ropes edged in turn with birds' eye or moss stitch (104). The tree motif, first seen at Eyemouth, is also used in many forms and can be alternated with diamonds to make an ornate and very beautiful jersey (105). Patterns are not 'called' names in Fisherrow, where the verb 'to cry' is used, and they cry the small basket stitch, so often used to replace moss stitch, herring scales, although moss stitch itself is still called birds' eye, like the birds' e'en of Flamborough. Panels are also called phases and vary with the need to add stitches for a wider garment.

Fisher Walk day is a holiday but it is the other 364 days that have made Fisherrow what it is and given the people of today their pride in their own past. It is easy to eat fresh salmon sandwiches and discuss knitting patterns, but this is not what earned the fisherwomen their nickname of 'Amazons'. Like many

104 *Diamonds in boxes, a traditional Fisherrow pattern*

105 *A Fisherrow pattern*

106 *Scots fisherlads before 1914*

108 *Three Fife lads, one wearing flag and chevron pattern*

107 *Scottish flag pattern chart*

other Scottish fishwives, there was much to be done and many had to kilt up their skirts and wade out to the boats with their menfolk on their backs, so that the day's fishing could start with the men dry, for in those days the boats were without decks.

This was but a start for, when fishing was by line, each hook, 1000 per line, had to be baited and bait had to be gathered. Furthermore, the aim of bait is to catch fish, and the woman's work included selling the fish. This could mean a 4 am start to walk into Edinburgh station and get a special ticket to travel across to the Fife shore to sell to the housewives there. Creels laden with fish were not light, and many a fisherwoman would place stones in her creel on the way home to keep her balanced, because she was more used to heavy baskets than light. If she had time during the day to wait – for a train, for the men to return, for it to be time for seeking bait and reddin' the lines again for the next day's work – she would knit. As she walked, or as she sat at her door

109 *Chart for flag and chevron pattern*

chatting with neighbours with an eye on the bairn, there was always knitting to keep her busy.

The work was not just shared by men and women, but the children, as soon as they could walk, helped with bait gathering and with bringing armfuls of grass from the braes to lay between the hooks on the lines as they were placed baited in a flat wooden scull or three-sided wooden box, so that there would be no tangle as they were paid out from the boat.

Flag and rope patterns

Before the 1914 war many guernseys were banded, but this photograph (106) shows a group of lads wearing, beside three banded guernseys, one patterned with the Scottish flag. This pattern can be found in many forms and is also known as the kilt pleat, because of the way the edge of the flag curls over (107).

Flags come in many sizes and can be teamed with other patterns (108). Standing at back right, the fisherlad wears a chevron pattern divided by two flags worked as a repeat, and shows the wider necks of pre-1914 (109).

Other flag patterns (110) may be alternated with rib, or, as in Caister, may panel a section along with ropes. Ropes are seen in Fife round the turn of the century, but it was often said in Scotland that ropes wore more easily, the raised twisted section taking greater friction, and were therefore less economical. This is, however, possibly a theory held only by those who disliked knitting cable patterns, for every area seems to be able to contribute its own version, from two twisted stitches to heavy and wide print o' the hoof cables.

By the 1920s the neater neck, with buttons usually of smoked pearl, replaced the wider neck (111). The man on the left is wearing a pattern which can be seen in many parts of Scotland and in various forms. Although unusual, it is similar to the many banded patterns used all round the coast which may be worked in a variety of different ways from moss

110 *Chart for flag and rib pattern*

111 *Button necked guernseys from Fife*

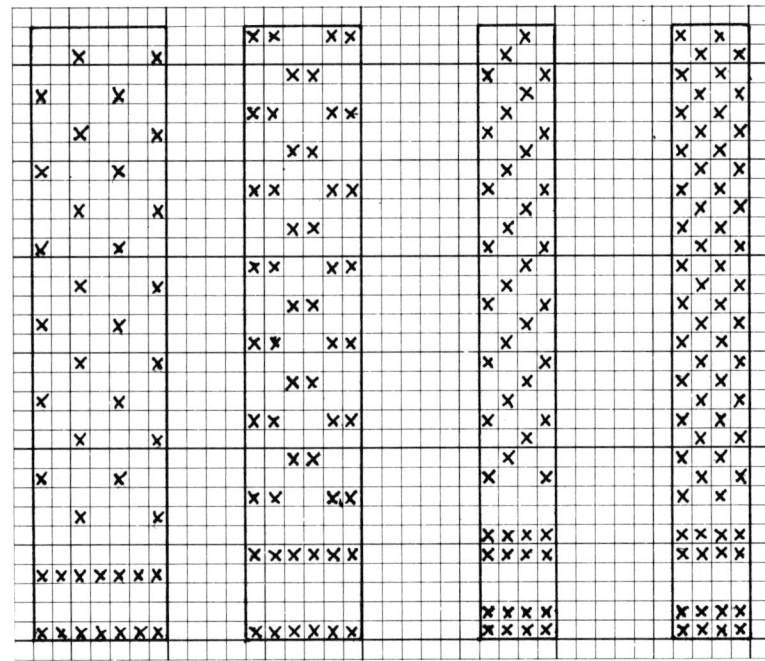

112 *Charts for banded patterns*

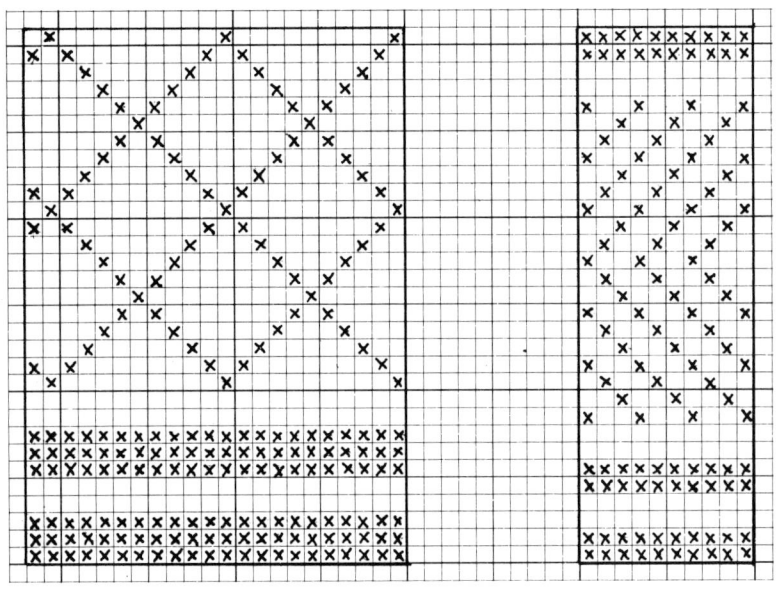

113 *Old 'brooks' on model in Scottish Fisheries Museum, Anstruther, found in 1975 after lying over 100 years, forgotten, in an attic*

stitch to wider diamonds, and which vary between single stitch lines and more noticeable lines worked over two stitches (112).

Although Poppleton's wool is ideal for all guernsey knitting, the wool used in Scotland before the 1939–45 war was very different. Most of it was supplied by Laidlaw's of Keith and it was softer, very smooth and round, so dark that it was almost black, and incredibly colour-fast, seemingly impervious to washing soda and saltwater alike, retaining its colour through years of hard wear. It knitted to a different tension from the yarn today and is responsible for some of the patterns looking slightly altered.

Really old guernseys are difficult to find in Scotland, partly because they were worn until they were finished, often having new welts knitted on and even sleeves or part sleeves replaced. They were also not considered to have any aesthetic or didactic value and so were not folded away for knitters of the future to find. When they were actually discarded, and not just rattled down to help mend another less-worn garment, they would be parcelled up with other waste wool and sent off to Laidlaw's in Keith as part payment for a new blanket. The scrap yarn was recycled at the mill and used in bothy blanket weaving.

There are times, however, when an old cupboard or forgotten kist (chest) is opened to reveal a hoard of

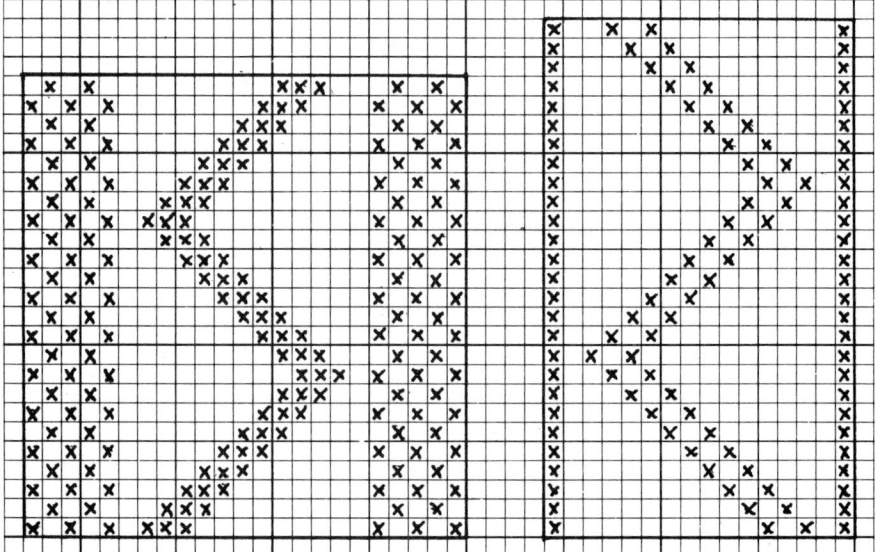

114 *Chart for zigzag lines and marriage lines*

old clothing. In 1975 some old 'brooks', as fishermen's clothes were called, were found in an attic where they had lain since 1860. Today, still showing signs of fish scales, they can be seen on a model fisherman in the Scottish Fisheries Museum at Anstruther (113). The guernsey on the model shows a zigzag pattern (114), called the multitude on the Northumberland coast, but more commonly known in a two-line version which is called marriage lines, and a single line which is likened to cliff paths (115).

Guernseys were not the only knitted garments worn by the fishermen, and in Anstruther can be found well-shaped drawers, with neat garter stitch opening and waistband (116), which were knitted in a pastel-shaded yarn sold specially and known as drawer worset as opposed to guernsey worset. Both were sold in cuts, as opposed to the knots that were usual in England. Wassit simply meant worsted, and the draper knew just which yarn to supply if he was told for what garment the knitter wanted it.

Slanting line patterns abound, but the version to be found in Cellardyke in Fife is unusual in its garter stitch background (117). Worked in the soft, nearly black wool used here and up the north-east coast, it

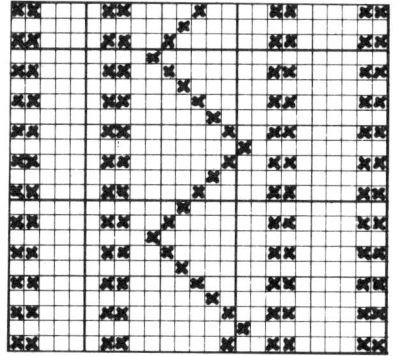

115 *Single line pattern chart, often called waves*

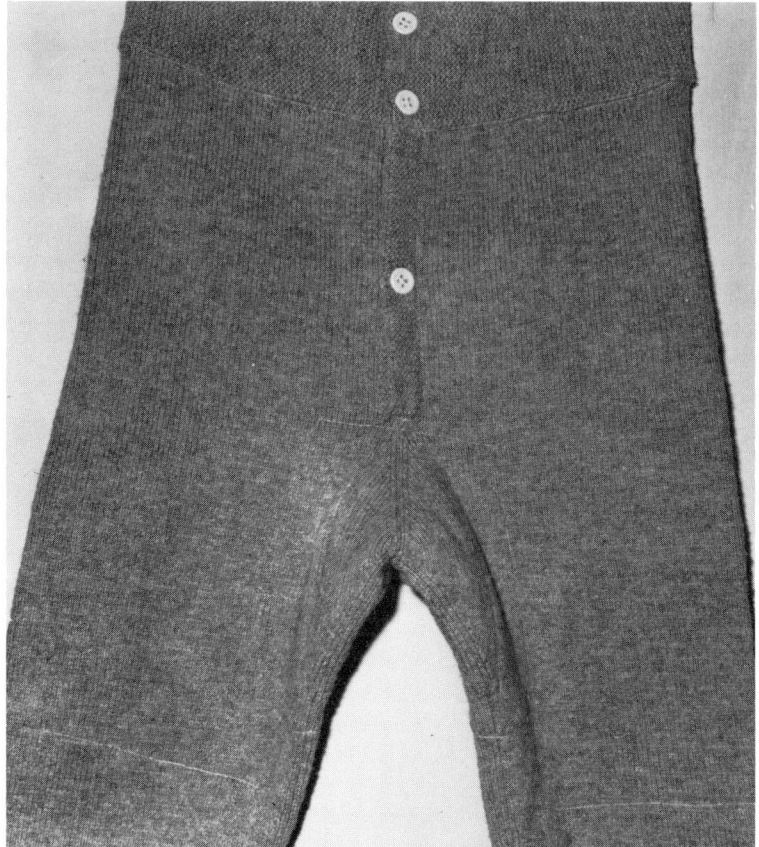

116 *Woollen handknitted drawers, showing the neat shaping*

117 *Chart for diagonal line and garter stitch pattern from Cellardyke*

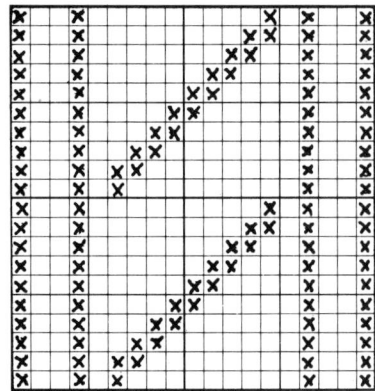

118 *Chart for diagonal line from Arbroath*

119 *'Tanker', otherwise unnamed, but wearing a guernsey with line pattern and herringbones barely discernible round the sleeves*

made a fine guernsey into which was usually tucked around the neck a dark Indian silk scarf.

Near Arbroath, the line slants the other way (118), while in Aberdeenshire an old, barely discernible picture shows 'Tanker', (119) otherwise un-named, but wearing a fine lined guernsey (120). The pattern was often to be found around the east coast and the Morayshire coast (121).

From Northumberland right up the coast a pattern known as church windows had been spoken of, but it was in the Arbuthnot Museum in Peterhead that an example at last materialised (122). It is easy to follow and in this version is divided by a small, almost dainty, cable which seems the perfect partner for the diagonal lines (123).

It is difficult to know where the east coast ends and the north begins, but for this book Peterhead is taken as the dividing line. Like many borders it is an unfair decision, because the patterns found in Aberdeenshire can be found in Morayshire and vice versa, but it has a certain convenience.

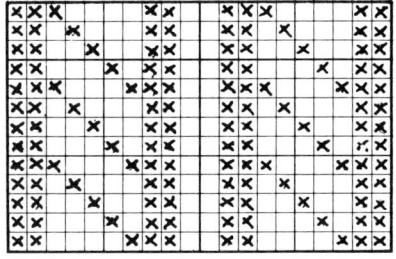

120 *'Tanker's pattern chart*

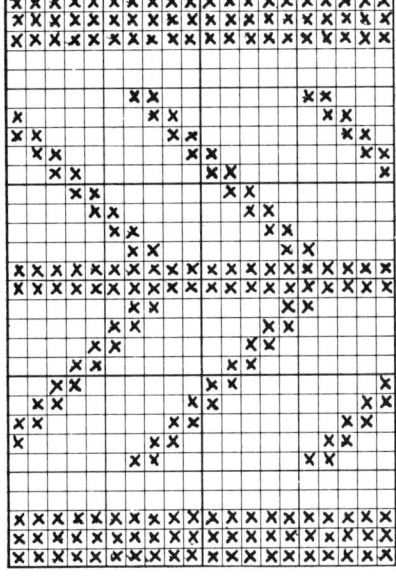

121 *Chart for herringbone pattern*

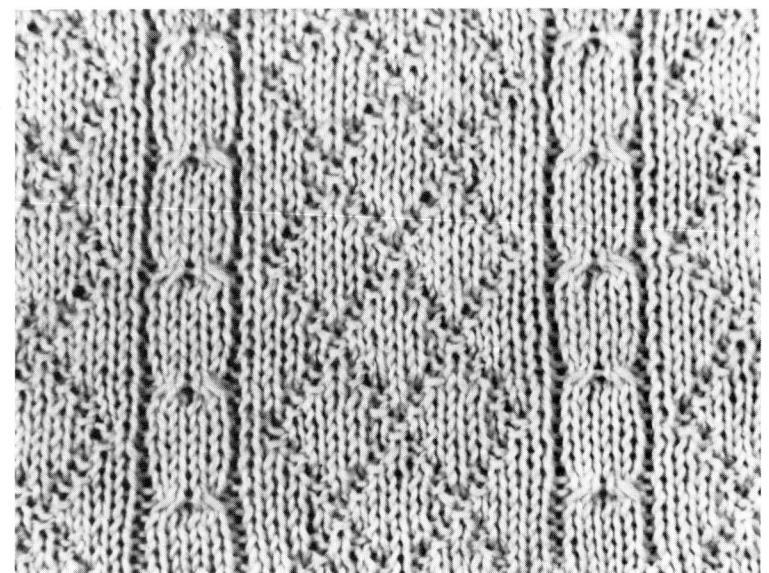

122 *Church windows pattern*

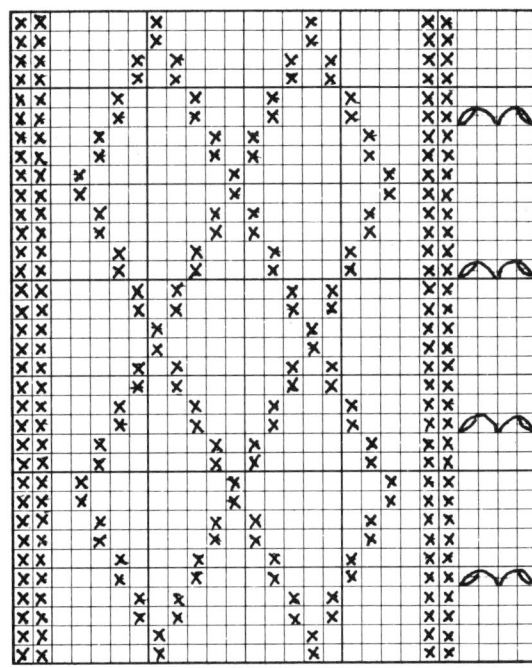

123 *Chart for church windows pattern*

8 North and West Scotland

It is hardly surprising to find that a major wool spinner, once setting out to record the patterns used traditionally in Aberdeenshire and Morayshire, found that every design was individual and that there was no basic pattern. All knitters have their own way of placing seam stitches and the variety of ways in which the seam stitches swell out to form the gusset are too many to mention. It is important to note that in every case they are part of the pattern, and in many cases are shallower than the gussets of the south, but are invariably outlined by a purl stitch. It is in Aberdeenshire and Morayshire that moss stitch becomes known as cat's teeth and basket stitch is no longer that, nor herring scales, but box pattern or even in Hopeman Mary-Ann's stitch.

The subject of names of patterns cropped up in Hopeman, and Mary More, then 86 and still able to knit a fine pair of socks, although she claimed they would no longer win prizes, laughed and laughed until she had to wipe her eyes when she heard that she was renowned for her pattern called the Road to Duffus. It was some time before she explained: 'You're like a' the rest. Always wanting to know the names. One day I was asked what that zigzag was called and I just said the Road to Duffus because it turns corners this way and that like the pattern.'

Mary More is like many another knitter. Left early a widow with young girls to bring up, she cleaned and worked at what she could and also mended nets and knitted guernseys and socks. Like many other professional knitters, she was also paid just as little as people felt they could decently get away with, not

124 *Two Portsoy youngsters*

125 *Mr Stewart in slipped zigzag pattern*

taking into consideration, perhaps not even realising, the amount of work that there was in one gansey.

She well remembered the day when one ordered gansey was only a few inches off being completed, when the gentleman it was for knocked on her door. Off for an interview, he wanted to wear his new guernsey. So the stitches were quickly slipped onto a thread and the man pulled it on, hiding the unfinished sleeve with his jacket. Away he went to the interview to return later for his gansey to be finished off properly.

Time and talk are both essential when seeking old patterns, and it is not always easy to direct the conversation without putting words into people's minds, or gaining agreement without actual remembrance on the part of the person supposedly recalling what they know. In Portsoy there had been a long and interesting conversation with Mrs Sutherland, also in her 80s but bright, busy and alert. Leaving the room to pick up a coat, words drifted out of the door as she completed the conversation to herself, 'Of course in my mother's time she'd have called them jersey-frocks'. It was the only time that the phrase was used in Scotland, but 50 or 60 years ago, just as in Cornwall, the guernsey was known as a frock or jersey-frock.

All along the small, clean fishing villages of the Morayshire coast, just as in Aberdeenshire, the number of variations is too great to make recording sense, for not every pattern is as simple as the chevron worn by one of the young lads from Portsoy (124). Many a prize winner is to be found in this area and many a knitter in Ayrshire or Argyll, in Wick or Eyemouth or Fife came originally from one or other side of Peterhead, learned from someone who came from this area or grew up here, before marrying a fisherman from elsewhere.

In communities where many people were of the same family and the surname is found again and again, identification is by bye-names. These extend to the women, although they often get their given name tagged onto their husband's bye-name. The reasons for the bye-names would form a book on their own and are often not what they seem to be on first hearing. Teapot and Snouts, Peenda and Cunner seem fairly distinguishing, but some like Daldon crop up in many places and almost need another bye-name to make identification exact.

In Lossiemouth Mr Stewart nobly turned model to show just a few of the many patterns that his mother, Mrs Isabella Stewart, laid out as an example of her knitting. Well into her 70s, her wires seem as busy as ever and she laid out for examination hearts and diamonds, ropes and twists, cat's teeth and a zigzag pattern she called waves, not unlike Mary More's Road to Duffus (125).

A favourite pattern was the old heart design, worked just as Carol Walkington in Bridlington had done (see final chapter), but panelled with a rope edged with basket, box or Mary-Ann's stitch (126).

126 *Mrs Stewart's heart and rope pattern chart*

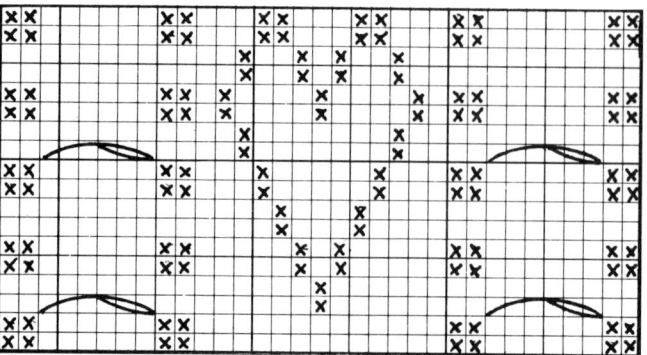

The cable twists to the right and is worked over six stitches. There is never a time when the guernsey knitter is too old to pay attention to detail, and the cuffs of this pattern were not made of ribbing but used a smaller version of the rope and box stitch, giving an interesting and unusual finish.

Mrs Stewart likes plain sleeves and her son appreciates ribbed necklines without buttons, particularly when it comes to wearing so many guernseys on one afternoon (127). The tree and tiny twice-repeated rope panel are typical of east coast patterns, but again the attention to detail shows in the neat shoulder strap chosen to suit the rest of the pattern. This shows box stitch, but the previous pattern had a diagonal line and other patterns showed the same small differences that speak of caring fingers (128).

The triangle, again with a rope and seed stitch outline, was a change from diamonds, recalling Cornwall in 1860 and Norfolk in 1900 (129).

It was of such care for detail that James Slater of Portsoy wrote:

'Deft work with the needles for pattern desired,
Silently counting the loops as required;
Loving thoughts of a dear one in the work intertwined,
Who will wear the planned vesture expertly designed . . .
. . . Knitting commercial we may cheaper obtain,
But there are values to treasure above earthly gain;
There is class and distinction in garments home-made,
Expertise and art that are lost in the "trade".'

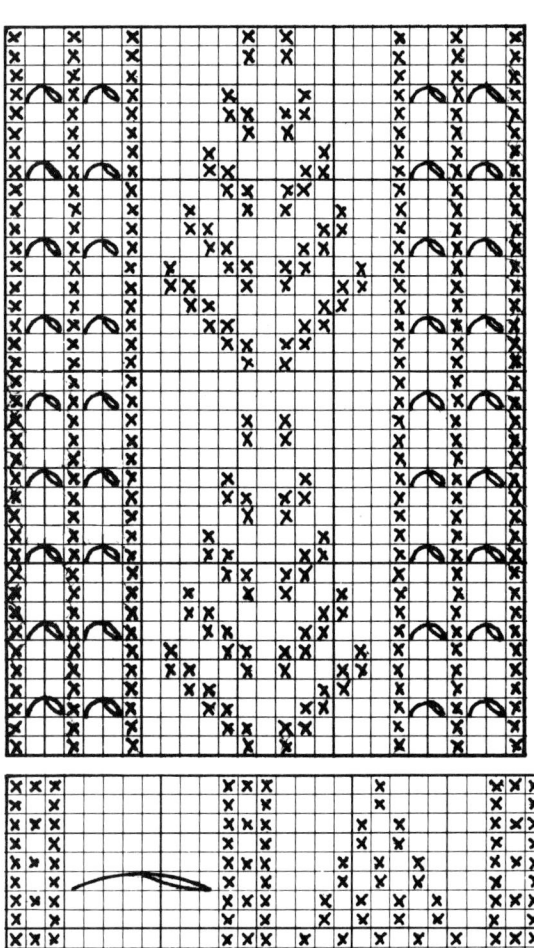
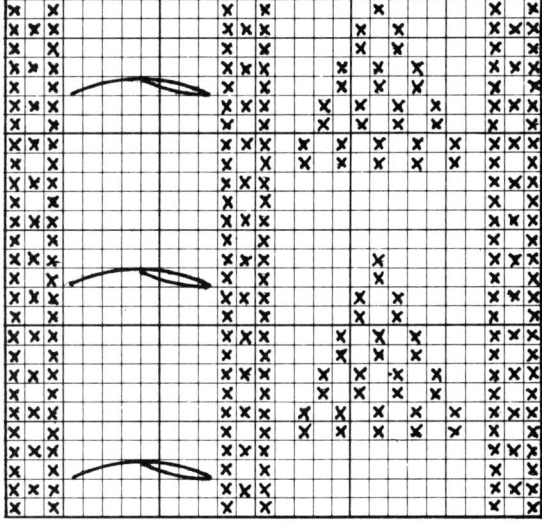

127 *Mr Stewart wearing the tree and miniature twin rope pattern*

128 *Chart for tree and rope pattern*

129 *Chart for half diamond and rope pattern*

Jersey patterns are not found only in the south and east of Scotland. From Galloway north to Cape Wrath, from Inverness to John O'Groats, variations of all the noted patterns have been used. From the north coast, particularly from Thurso, come patterns which are complex and rich in texture, and which have an interesting pattern-related shoulder strap.

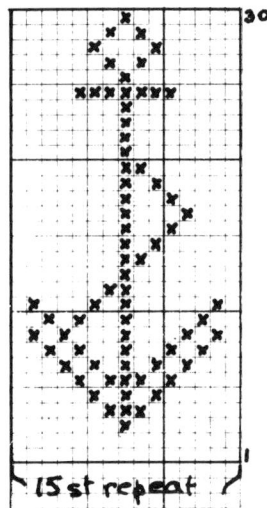

130 *Angus Mackay's anchor pattern chart*

131 *Close up of Angus Mackay's pattern*

132 *Skipper Donald Thomson and cup winning crew of* The Children's Friend, *at the Thurso Regatta, 1912*

Angus Mackay's decorative gansey, (130) is of interest because it shows an anchor pattern, often recorded but seldom seen in photographs (131). It is also banded, more like the patterned ganseys of Eriskay, with a motif not unlike the starfish still used today. Different also is the central cable because it is the seldom seen print o' the hoof cable, found first in the Filey fishermen's choir patterns.

This formal picture (132) shows the skipper Donald Thomson and the crew of *The Children's Friend* – the McKenzie Cup winners of the Thurso 1912 Gala. Donald Thomson wears a pattern that is not unlike the one worn in Polperro by Jim Curtis as early as 1860, although one has ropes and the other has not, and it is not the only pattern which seems surprisingly like the early Cornish patterns (133). The close up shows a purled diamond but this pattern can also be worked with a moss stitch diamond. The working instructions for it are in the final chapter.

133 *Close up of Donald Thomson's pattern*

134 *George Reid's pattern*

135 *George Reid's pattern chart*

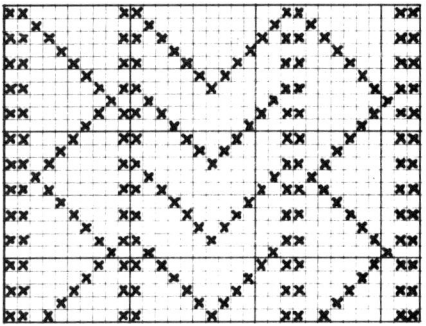

The pattern worn by George Reid, back row right (134), is composed of waves and chevrons, each vying with the other for predominance (135). The shoulder straps of these ganseys are of special note and can be seen close up (28). The instructions can be found in Donald Thomson's pattern.

Another pattern (136) which, although less ornate, is still bold and more textured than might at first be realised, because of its tendency to curl or pleat, is the Caithness flag pattern (137).

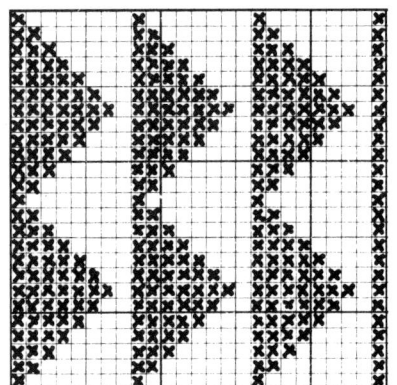

136 *and* **137** *Caithness flag pattern and chart*

138 *Front cover pattern chart – main pattern*

ERISKAY

It seems that there is some reality in suggesting that the more remote the place and the less daily contact had between the people and other places, the more imaginative are the patterns worked by the knitters. The cover pattern is from Eriskay in the past, but today the women working in a cooperative, with their ganseys being sold in London and abroad, produce just as interesting patterns, one of which is given in the final chapter.

The front cover is typical of the use of three distinct patterns over three areas. The chart in figure

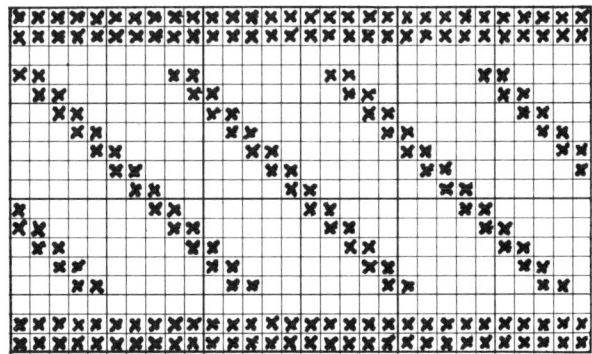

139 *Front cover pattern chart – diagonal line band*

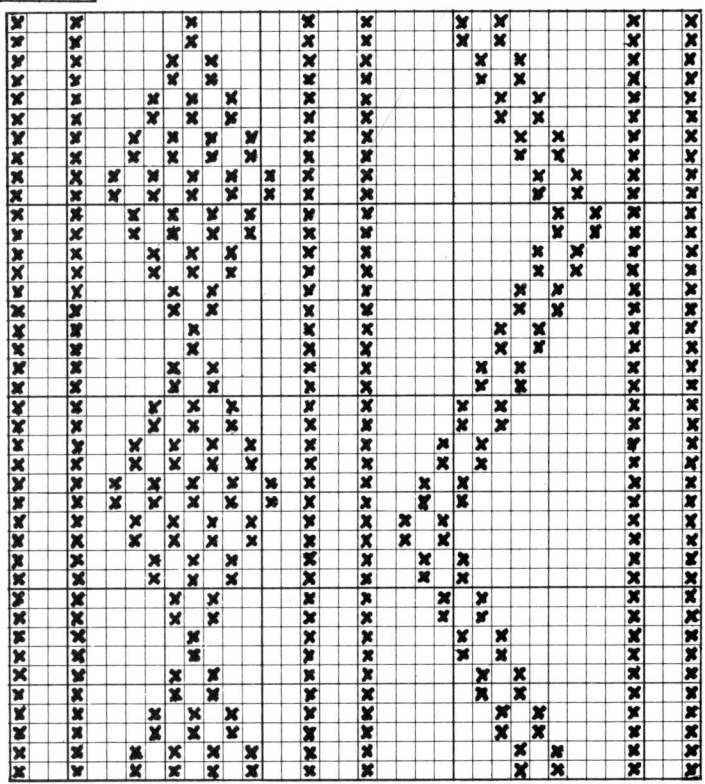

140 *Front cover pattern chart – panel repeats*

138 shows the pattern that might be called the main pattern, which is worked on the top section of front and back and which separates the main motifs by a double chevron panel. Figure 139 shows the diagonal line band that divides the top and lower sections, and 140 shows the panel repeats for the lower section.

The close up picture shows the tree with the starfish (141) above it, openwork panels at the side and bands between the motifs of Indian corn stitch (142). Possibly the pattern shown on the lower section of a garment bought commercially is of even

141 *Close up of starfish and tree pattern on an Eriskay guernsey, 1979*

142 *Chart for starfish pattern*

more interest because it shows the pattern used on the cover design, photographed 30 to 50 years earlier (143). There is a slight difference in the diamond but the wave remains the same.

Indian corn stitch

An adaptable stitch, this can be used singly worked with only two stitches to give the appearance of tying them together, or it can be worked as an allover design as seen on the shoulder strap. It can also be used for any multiple of two and is easily added as dividing bars, as if to underline another pattern. Worked as a pattern it takes an even number of stitches with two end stitches.

First row K1, * yon, K2, with left needle tip lift yon over both the K2 stitches and off needle point so that it lies across the two stitches, horizontally, rep from * to last st, K1.
Second, third and fourth rows st st.
  These four rows are the usual repeat. It can, however, be worked with as many rows between the barred rows as required.

SHETLAND

The real tradition of Shetland knitters is their coloured work, but, despite the beliefs of some, guernseys were not unknown nor unknitted on the islands. The photograph (144), taken about 1900, shows the crew of the *Willie Bruce* and shows two of the banded guernseys. Unfortunately the other patterns are not clear. From left to right the

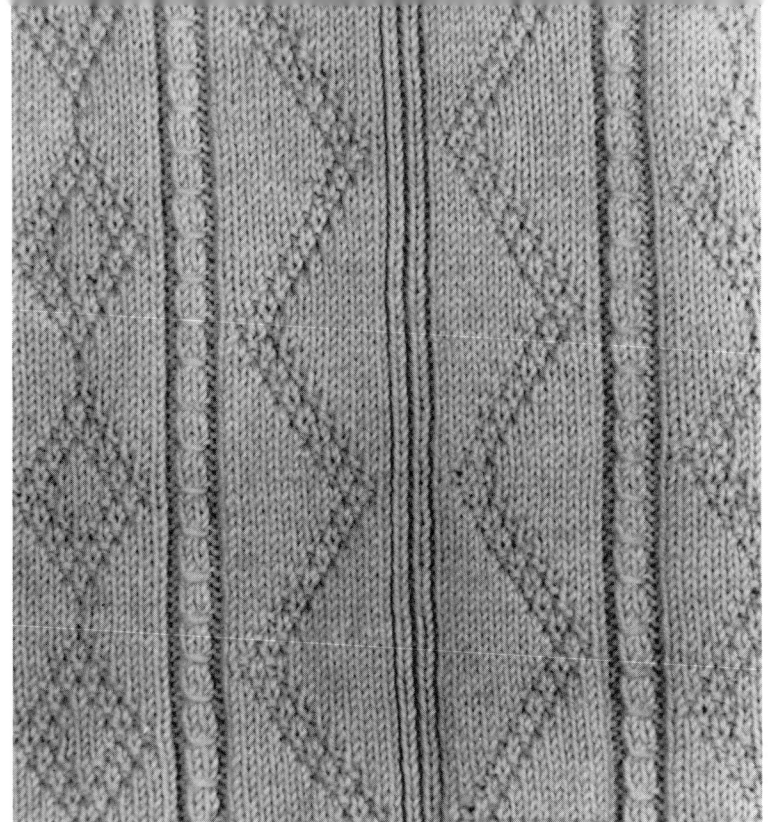

143 *Eriskay diamond and wave pattern*

members of the crew are George Sutherland, Eddie Polson, Robbie Bruce, Hughie Sutherland, Magnie Polson, Johnny Polson and Lowrie Bruce.

In Whalsay more patterned guernseys were also worked using diamonds and the larger, more traditional, motifs, possibly influenced by fishermen visiting the island from Holland and Denmark where similar patterns seem to have been used.

144 *Crew of the* Willie Bruce, *Shetland*

9 Designing a guernsey

The ideal way to use patterns like these is to experiment with small swatches. When a pattern particularly pleases you, or when you make a new combination that you like, make a guernsey to your own measurements for one of your family.

Starting is the most difficult point but, with this list of questions which must be answered and the following table as an approximate guide, beginning should be made easy. No approximate measurements can ever take the place of your own exact measurements, but they can serve as a starting point if you are a newcomer to designing.

DECISIONS

You will need to know the answer to all these questions, so decide before you start your guernsey and make a note of your decisions. It is no use working a ribbed welt without thinking beyond. There are points that you need to know right from the cast on edge, such as whether you are to have seam stitches or panels. You may want them to form naturally out of the ribbing and will be stopped from doing this if you have not cast on a number of stitches that will allow for this.

Which thickness of yarn are you going to use?
What is your tension?
Who is it for and what is the size?
How long from shoulder to lower edge?
How long from lower edge to underarm?
(The difference between the last two numbers will give the depth of armhole.)
How deep is the welt to be?

What type of rib or pattern will be used for the welt?
How many stitches are to be in the gusset?
How many rows or rounds between each increase?
(With this information you can tell the gusset length and subtracted from the total side length this will give you the starting point for the gusset.)
Where is the pattern to start?
Does the pattern draw in the material?
(Ropes will need more stitches for same width, moss stitch will give the greatest width. Vertical patterns that have many panels divided by ribs will tend to draw in like ribbing, banded designs may need extra rows to gain their full length.)
What type of shoulder finishing are you going to use?
What type of neck finishing are you going to use?
Is the sleeve to be banded or plain?
How deep is the cuff to be?
Have you remembered seam stitches?

Tension

The number of stitches to 1 in (2.5 cm) or to 4 in (10 cm) is of importance even if you are using a written pattern with all these decisions made for you. It is also important to know how many rows or rounds make the same measurement. It is essential to work a swatch in stocking stitch and also in the pattern you are going to use. The larger the swatch, the more accurate your measuring of it will be. It is not possible to measure 1 in (2.5 cm) on a tiny square and be certain that you have discovered whether there are fractions of stitches involved. Measure over not less than 4 in (10 cm) on a swatch that is at least several stitches larger: twice as large is ideal. Measure with the swatch flat and pinned in place to a board, if it is likely to curl and make measuring difficult.

DIAGRAMS

Make a small diagram giving the measurements. This will serve two purposes: it means that you have made the decisions you will need to know and it will show up anything that you have forgotten to measure.

Graph paper

It is easier to chart on graph paper what the pattern is going to do than to visualise it in your mind. On graph paper every stitch will be represented and must be accounted for. Do remember, however, that although graph paper will give you a map to find your way from the start of the pattern to the end, it will not be in scale and, knitted up, it will become more squat, changing the shapes you have placed on it. A knitted stitch is wider than it is high, while a graph paper shape is square.

Using the chart
The chart shows the chest sizes on the first line, giving length of side and armhole on the lines below. The depth of the gusset is measured by the number of increases worked every four rounds, and all the stitches and rows are based on using Poppleton's 5-ply Guernsey at a tension of 28 sts to 4 in (10 cm).

The number of stitches to cast on for the welt is fewer than the suggested number for the total width. A well-fitting ribbed welt is thus made, with the additional stitches increased on the last round (half

145 *Coastguard and lifeboat crew, Robin Hood's Bay*

146 *Scarborough fisherman with basket stitch yoke*

the number on the front and half on the back). The total number of stitches, unless you are using many ropes or a complex pattern, could be expected to lie between the figures given, a tighter fit being given by the smaller number.

Pattern
The photograph of the coastguard and lifeboat crews of Robin Hood's Bay shows some interesting patterns, but the one second from the left on the front row is a good design to start with (145). Yoked, it is patterned in moss stitch and can also be worked in basket, box or Mary-Ann's stitch, like this version worn by a Scarborough fisherman (146).

A banded yoke is the next variation like the one worn by Harry Freeman (3). If you think it is going to be too difficult to fit the bands into the exact number of rows, there is no need to discard the idea. Many knitters in the past simply repeated the band until the work was the required length, not minding

Guidance Chart

Chest size	24	26	28	30	32	34	36	38	40	42	44	46	in
	62	67	72	77	82	87	92	97	102	107	112	117	cm
Length to shoulder	14	16	18	20	$21\frac{3}{4}$	24	25	$25\frac{1}{2}$	$26\frac{1}{4}$	$26\frac{1}{2}$	$27\frac{1}{4}$	28	in
	36	41	46	51	56	61	63	65	67	68	70	71	cm
Depth of armhole	$6\frac{1}{4}$	7	$7\frac{1}{2}$	8	$8\frac{1}{2}$	9	$9\frac{1}{4}$	$9\frac{1}{2}$	$9\frac{3}{4}$	10	$10\frac{1}{2}$	11	in
	16	18	19	21	22	23	23	24	25	26	27	28	cm
Length of side	$7\frac{3}{4}$	9	$10\frac{1}{2}$	12	$13\frac{1}{4}$	15	$15\frac{3}{4}$	16	$16\frac{1}{2}$	$16\frac{1}{2}$	17	17	in
	20	23	27	30	34	38	40	41	42	42	43	43	cm
Length to gusset	$5\frac{1}{2}$	$6\frac{1}{2}$	8	$8\frac{1}{2}$	10	$11\frac{3}{4}$	$12\frac{1}{2}$	$12\frac{1}{4}$	$12\frac{1}{2}$	$12\frac{1}{2}$	$12\frac{1}{2}$	$12\frac{1}{2}$	in
	14	17	21	22	26	30	32	31	32	32	32	32	cm
No. of sts in gusset (inc 2 sts every 4 rounds)	12	12	12	16	16	16	16	20	20	20	24	24	
No. of sts for welt	172–180	184–192	200–208	208–220	222–234	236–248	252–268	260–276	276–288	288–300	296–312	312–328	
No. of sts to inc for body	12	12	12	16	16	16	16	20	20	20	24	24	
No. of sts for body	184–192	196–204	212–220	224–236	238–250	252–264	268–280	280–296	296–308	308–320	320–336	336–352	

in the least if the neck came just after the start of a new band.

The patterns on chart 131 will give you a suggestion for fillings, or you can think out your own. A different filling in each band makes the knitting all the more interesting.

Tips
Look at many of the photographs and you will see that knitters have solved some of the snares for you. Starting a pattern immediately after the rib can give you problems in placing the seam stitch and knowing just where to increase so that the pattern is not in conflict with a ribbed welt. This gives no problem if you work a plain section immediately above the welt before beginning the pattern. Originally worked so that a thread could be pulled and loops picked up for working a new welt if the old one became worn, it serves as a very good division between one area and the next. If you are determined to start the pattern immediately, another point that makes it easier to work is to make two or three rounds of garter stitch, or two rounds of Indian corn stitch as a frame to the pattern (96).

Sizes
Guernseys are supposed to fit reasonably neatly, although many of the photographs raise doubts about this fact. If you know that you want a loose-fitting garment, choose a size larger; if you want a small one, consider a size smaller.

10 Guernsey designs

A step towards designing for yourself is to knit from instructions that are already worked out for you. These designs are all based on traditional garments and are altered only if it was considered that a repeat might be easier or if sizes made it impossible to use the exact number of stitches shown in the photographs. They represent one of each of the main styles of stitch types.

SIZES
All sizes are given with today's finished garment in mind. Jerseys are now usually knitted to a longer length than was common in the past, when many stopped at waist-length. The measurements given are average, and if you want to add to the length read through the pattern before you begin to knit so that you know the best point at which length can be added without affecting the patterns.

APPLEDORE JERSEY
Although this is a fine jersey, worked in 4-ply not 5-ply, it is a good introduction to jersey knitting because it is plain with patterning only on the shoulders. The pattern is worked on the front shoulders and knitted together with the same number of stitches from the back on the wrong side, making an almost invisible finish (148).

Materials
11[11:12:13] 50g balls of Poppleton's 4-ply Guernsey
1 set of $2\frac{1}{4}$mm (No. 13) needles (use 2 sets if short needles are used) or 1 set of $2\frac{1}{4}$mm needles and 1 circular $2\frac{1}{4}$mm needle for the body

147 Appledore jersey with pattern shoulder strap

148 *Shoulder strap detail*

Measurements
To fit a 36[38:40:42] in chest (92[97:102:107] cm)
Length to shoulder, 26½[27:27½:28] in (68[69:70:71] cm)
Sleeve seam, 20[20:21:22] in (51[51:53:56] cm)

Tension
36 sts and 44 rows or rounds to 4 in (10 cm) over st st worked on 2¼mm needles

Instructions
Begin at lower edge of body.
With 2¼mm needles or circular needle cast on 352[368:384:400] sts.
Join into a circle taking care not to twist sts. Place a loop of contrast yarn before the 1st st to mark the beginning of the round.
Work in rounds of K1, P1 rib for 2½ in (4 cm).
Work 6 rounds P.

Change to st st and place seam stitches
1st round * K176[184:192:200], M1 purlwise, rep from * once.
2nd round * 176[184:192:200], P1, rep from * once.
Rep 2nd round until work measures 14½ in (34 cm) or required length to start of underarm gusset, approx

12[12½:13:13½] in (31[32:33:34] cm) less than finished length. Length should be adjusted at this point, working fewer or more rounds as required.

Begin gusset
1st round * K176[184:192:200], (P1, K1, K1b, P1) all in next st, rep from * once.
2nd round * K176[184:192:200], P1, K2, P1, rep from * once.
Rep 2nd round twice.
5th round * K176[184:192:200], P1, M1, K2, M1, P1, rep from * once.
6th round * K176[184:192:200], P1, K4, P1, rep from * once.
Rep 6th round twice.
Cont in this way, inc 1 st at each side of gusset on every 4th round until there are 40 sts without counting seams sts.
Work 1 round more.

Divide for front and back
1st round * K176[184:192:200], P1, K40, P1, slip these last 42 sts onto holder, rep from * once.
** Complete front on next 176[184:192:200] sts, leaving rem sts unworked. Work in rows until armhole measures 4½[5:5½:6] in (12[13:14:15] cm), ending with a P row. **

Right shoulder
1st row K120[124:132:136] sts, leave these sts on holder and K rem 56[60:60:64] sts for right shoulder strap.
*** *2nd row* (wrong side facing) K56[60:60:64].
3rd row P56[60:60:64].
Rep 2nd and 3rd rows once.
6th row P56[60:60:64].
7th and 8th rows * K2, P2, rep from * to end.
9th and 10th rows * P2, K2, rep from * to end.
Rep 7th–10th rows 4 times.
27th row K56[60:60:64].
28th–31st rows rep as for 2nd and 3rd rows twice.
32nd row P.
33rd row K.
Rep from 2nd–31st row once. Leave these sts on holder. ***

Left shoulder
With wrong side of rem 120[124:132:136] sts of front facing, slip centre 64[64:72:72] sts to holder for neck, rejoin yarn to rem 56[60:60:64] sts and work as for other shoulder from *** to ***.

Back
With right side of rem sts facing, slip 42 sts for both gussets onto holders, leaving 176[184:192:200] sts for back. Rejoin yarn to these sts with right side facing and work as for front from ** to **.

To join shoulders
With wrong side of back facing, hold needle with 56[60:60:64] sts of right front shoulder behind back, right side of front touching right side of back. Cast off 56[60:60:64] sts together by taking 1 st from each needle and knitting both together, lifting 1st over 2nd every time 2 sts are on right needle in the usual way. Work other shoulder in same way, leaving centre 64[64:72:72] sts on holder for neck.

Neck
With set of $2\frac{1}{4}$mm needles and right side of work facing, K64[64:72:72] sts from back neck, K up 30 sts from neck edge of shoulder strap, K64[64:72:72] sts from front neck and K up 30 sts from neck edge. Arrange these 188[188:204:204] sts evenly on 3 needles. Work in rounds of K1, P1 rib for 6 in (15 cm). Cast off loosely in rib.

Sleeves
With set of $2\frac{1}{4}$mm needles and right side of armhole facing, K up 40[44:48:52] sts up side of armhole, K up 30 sts from shoulder strap edge, K up 40[44:48:52] sts from other side of armhole and P1, K40, P1 across gusset sts.
1st round K110[118:126:134], P1, K40, P1.
Rep 1st round once.
3rd round K110[118:126:134], P1, K2 tog tbl, K36, K2 tog, P1.
4th round K110[118:126:134], P1, K38, P1.
Cont in this way, dec 1 st at each side of gusset until 2 sts rem on gusset. Work 3 rounds.
Next round K110[118:126:134], P4 tog.
Next round K110[118:126:134], P1.
Rep last round twice.

Next round K2 tog tbl, K to last 3 sts, K2 tog, P1.
Next round K to last st, P1.
Rep last round twice.
Cont dec 1 st at each side of seam st on every 4th round until 72 sts rem.
Work until sleeve measures 6 in (15 cm) less than required length.
Cont in rounds of K1, P1 rib for 6 in (15 cm).
Cast off loosely in rib.
Work other sleeve in same way.

To complete
Darn in all ends. Press on wrong side under a damp cloth with a warm iron, omitting ribbed and patterned areas.

BRIDLINGTON GANSEY
Hearts, diamonds and ropes make this gansey suitable for any seafaring family. The pattern given is for the children's sizes but you can add to the moss stitch panels or enlarge the diamonds for adult sizes, and work one for all the family just as Carol did for Craig (149), Grant and Fred, who is coxswain of Bridlington's life-boat.

Materials
8[9:10] 50g balls of Poppleton's 5-ply Guernsey
1 set of $2\frac{3}{4}$mm (No. 12) needles or 1 set of $2\frac{3}{4}$mm needles and one circular $2\frac{3}{4}$mm needle. (If you use a set of needles for the body you may need to use 2 sets if the needles are short)
I cable needle

Measurements
To fit chest 26[28:30] in (62[67:72] cm)
Length to shoulder $14\frac{1}{2}$[16:18] in (36[41:46] cm)
Sleeve seam 11[$12\frac{1}{2}$:14] in (28[32:36] cm)

Tension
28 sts and 38 rows to 4 in (10 cm) over st st on $2\frac{3}{4}$mm needles

Special abbreviations
C4   sl next 2 sts to CN, hold at back, K2, K2 sts from CN.
C6   sl next 3 sts to CN, hold at back, K3, K3 from CN.

149 *Grant and Craig Walkington in their heart patterned ganseys*

Instructions

Begin at lower edge.
With set of 2¾mm needles or circular needle and double yarn cast on 208[224:240] sts, using 3 strand cast on.
1st round (using 2 strands of yarn), * P1, K2, P1, rep from * to end.
Place marker loops before 1st and after 104th[112th:120th] sts to mark sides of front and back.
Work 5 rounds more in rib as set with yarn double, then work with 1 strand only until rib measures 2 in (5 cm), inc 10 sts evenly on last round, 5 sts between 1st and 2nd marker and 5 sts between 2nd marker and end of round. 218[234:250] sts.

Place seam sts
Next round * P1, K107[115:123], P1, rep from * once. Rep this round until work measures 3[3½:4] in (8[9:10] cm) from cast on edge.

Begin pattern
1st round * P1, (K1, P1) 3 times, K4 [6:6], P1, (K1, P1) 3 times, work 1st row of diamond from chart 150a [a:b] over next 11[11:15] sts, P1, (K1, P1) 3 times, K4[6:6], P1, (K1, P1) 3 times, work 1st row of heart from the chart in 151 over next 15 sts, P1, (K1, P1) 3 times, K4[6:6], P1, (K1, P1) 3 times, work 1st row of diamond from chart 150a[a:b] over next 11[11:15] sts, P1, (K1, P1) 3 times, K4[6:6], P1, (K1, P1) 3 times, rep from * once.
2nd round * (P1, K2) twice, P1, K4[6:6], P1, (K2, P1) twice, work 2nd row of diamond over next 11[11:15], P1, (K2, P1) twice, K4[6:6], P1, (K2, P1) twice, work 2nd row of heart over next 15, P1, (K2, P1) twice, K4[6:6], P1, (K2, P1) twice, work 2nd row of diamond over next 11[11:15], P1, (K2, P1) twice, K4[6:6], P1, (K2, P1) twice, rep from * once.
3rd round as 1st round, working 3rd row from charts.
4th round as 2nd round, working 4th row from charts.
5th round * P1, (K1, P1) 3 times, C4[C6:C6], P1, (K1, P1) 3 times, work diamond, P1, (K1, P1) 3 times, C4[C6:C6], P1, (K1, P1) 3 times, work heart, P1, (K1, P1) 3 times, C4[C6:C6], P1, (K1, P1) 3 times, work diamond, P1, (K1, P1) 3 times, C4[C6:C6], P1, (K1, P1) 3 times, rep from * once.

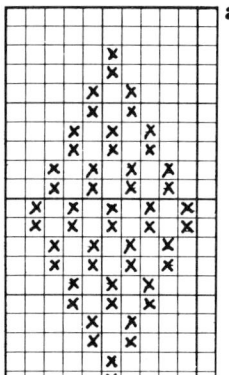

150 *Chart for diamonds*

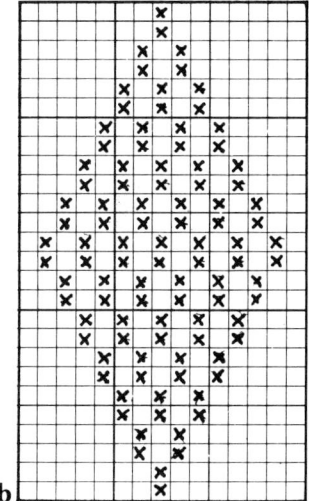

151 *Chart for hearts*

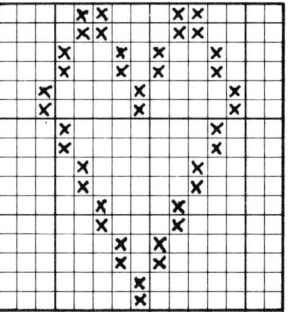

6th round as 2nd, working 6th row from charts.
Cont in this way keeping patts as set working cable twists on every 6th round and rep heart and diamond motifs as required.
Cont until work measures 7[8½:10] in (18[22:25] cm) or approximately 2 in (5 cm) less than required side length.
If a longer garment is required work extra rounds at this point.

Begin gusset
1st round slip 1st st (seam st) onto end of last needle and P it, * patt 107[115:123], P1, lift yarn before next seam st and make 2 sts by working knitwise into front, and back of loop, P1, rep from * once.
2nd round * patt 107[115:123], P1, K2, P1, rep from * once.
3rd round work as 2nd round.
4th round * patt 107[115:123], P1, K1, m2 as before, K1, P1, rep from * once.
Work 2 rounds keeping seam sts correct at each side of gusset.
7th round * patt 107[115:123], P1, K1, m1, K2, m1, K1, P1, rep from * once.
Cont in this way keeping seam sts and patt as set and inc 1 st at each side of gusset inside edge st on every 3rd round until there are 12 sts on the gusset.
Cont without shaping until 3 rounds more have been worked or until side length of 9[10½:12] in (23[27:30] cm) is reached or required side length.

Divide for armholes
1st row patt 107[115:123], turn. Complete back on these sts leaving rem sts on holder.

Back
Work in rows, keeping patt as set until work measures 13½[15:17] in (33.5[38.5:43.5] cm), ending with a wrong side row.
Leave sts on holder.

Front
With right side of work facing slip 12 gusset sts and both seam sts at each side onto a holder. With right side facing, rejoin yarn to rem 107[115:123] sts and work until same length as back.
Leave sts on holder. Do not break off yarn.

Left shoulder strap
1st row P35[38:41], turn. Complete strap on these sts, leaving rem sts on holder.
2nd row K.
3rd row K.
4th row P.
Rep 1st–4th rows 4 times.
Place these 35[38:41] sts against same number on left side of back with right sides touching and cast off together on wrong side by working 1 st from both front and back tog as 1 st, lifting 1 over the other each time there are 2 sts on the needle. Complete all the shoulder strap sts in this way.

Right shoulder strap
With right side of work facing, slip centre 37[39:41] sts onto holder for neckband. Rejoin yarn to rem 35[38:41] sts and work in same way as for other side, casting off back and front tog on wrong side.

Neckband
With set of $2\frac{3}{4}$mm needles and right side of work facing, K across 37[39:41] sts from centre back, K up 15 sts from shoulder strap K across 37[39:41] sts from centre front, K up 15 sts from shoulder strap. 104[108:112] sts.
Work in K2, P2 rib for $1\frac{1}{2}$ in (4 cm). Cast off in rib.

Sleeves
With set of $2\frac{3}{4}$mm needles and right side of left armhole facing, K up 35[39:42] sts from end of gusset to centre, K 1 st from centre, K up 35[39:42] sts from other side of armhole, (P1, K12, P1 across gusset sts).

Place pattern
1st round K10[12:15], P1, (K1, P1) 3 times, K4[6:6], P1, (K1, P1) 3 times. Work heart across next 15 sts starting at top of chart and working down, P1, (K1, P1) 3 times, K4[6:6], P1, (K1, P1) 3 times, K10[12:15], P1, K12, P1.
2nd round K10[12:15], P1, (K2, P1) twice, K4[6:6], P1, (K2, P1) twice. Work heart over next 15 sts, P1, (K2, P1) twice, K4[6:6], P1, (K2, P1) twice, K10[12:15], P1, K12, P1.
3rd round as 1st round to seam st, P1, K1, sl 1, K1, psso, K6, K2 tog, K1, P1.
4th round as 2nd round to seam st, P1, K10, P1.

5th round K10[12:15], P1, (K1, P1) 3 times, C4[C6:C6], P1, (K1, P1) 3 times, work heart over next 15 sts, P1, (K1, P1) 3 times, C4[C6:C6], P1, (K1, P1) 3 times, K10[12:15], P1, K10, P1.

Cont in this way dec 1 st inside edge st at each side of gusset on every 3rd round until only seam sts rem, then every 6th round keeping patt as set and working cable on every 6th round. Work until 4[5:5] hearts have been completed. Cont in st st, keeping seam sts as set and cont dec 1 st at each side of seam sts until 49[53:57] sts rem. Work without shaping until sleeve measures 9[$10\frac{1}{2}$:12] in (23[27:31] cm) or 2 in (5 cm) less than required finished length, dec 1 st on last round. Work in K2, P2 rib for 2 in (5 cm). Cast off in rib.

CAMPBELTOWN GUERNSEY

Panels of cabled ropes, with tailored lines of ribbing between, make this guernsey quite distinctive. The shoulder strap adds to the distinction with a basket stitch pattern in contrast to the vertical lines, and is completed by a neat straight line of cast off above the pattern on the right side of the work at the shoulder. The sleeves can be topped with more basket stitch, left plain or often have a band of herringbone worked about 4 in (10 cm) from the top. A ribbed welt has been added to this design.

Materials
14[15:16:17] 50g balls of Poppleton's 5-ply Guernsey
1 set of $2\frac{1}{4}$mm (No. 13) needles or 1 set and 1 circular needle
1 set of 3mm (No. 11) needles or 1 set and one circular needle
1 cable needle

Measurements
To fit chest 36[38:40:42] in (92[97:102:107] cm)
Length to shoulder $25\frac{1}{2}$[$26\frac{1}{2}$:$26\frac{3}{4}$:$27\frac{1}{2}$] in (65[67:68:70] cm)
Sleeve seam 18[18:19:20] in (46[46:49:51] cm)

Tension
26 sts and 36 rounds or rows to 4 in (10 cm) over st st on 3mm needles

Special abbreviation
C6 sl next 3 sts to CN, hold at front, K3, K3 from CN.

152 *Campbeltown guernsey*

Instructions
With set of 2¼mm needles or circular 2¼mm needle cast on 288[304:320:336] sts.
1st round * P1, K2, P1, rep from * to end.
Place marker loop before 1st st and after 144th[152nd:160th:168th] sts to mark sides.
Cont in rib as set until 2 in (5 cm).

Place seam sts
Next round * P1, K142[150:158:166], P1, rep from * once.
Rep last row until work measures 1¼ in (3 cm) from top of ribbing.
If preferred this stocking stitch section may be omitted and the pattern may start immediately after placing the seam sts.

Begin pattern and change to 3mm needles
1st round ** P1, (K2, P2) 3[4:5:6] times, * K6, P2, (K2, P2) 5 times, rep from * to last 18[22:26:30] sts before 2nd marker, K6, (P2, K2) 3[4:5:6] times, P1, rep from ** to end of round.
2nd round * P1, rib 12[16:20:24], K6, (rib 12, K6) 5 times, rib 12[16:20:24], P1, rep from * once.
3rd and 4th rounds work as for 2nd round.
5th round * P1, rib 12[16:20:24], C6, (rib 12, C6) 5 times, rib 12[16:20:24], P1, rep from * once.
6th, 7th and 8th rounds as 2nd round.
These 8 rounds form the patt and are rep throughout the body.
Cont until work measures 12[12½:12½:12¾] in (31[32:32:33] cm) or required side length less approximately 4 in (10 cm).
For additional length add rounds at this point.

Begin gusset
1st round * P1, m1, patt 142[150:158:166], m1, P1, rep from * once.
2nd and 3rd rounds * P1, K1, patt 142[150:158:166], K1, P1, rep from * once.
4th round * P1, m1, K1, patt 142[150:158:166], K1, m1, P1, rep from * once.
5th and 6th rounds * P1, K2, patt 142[150:158:166], K2, P1, rep from * once.
7th round * P1, m1, K2, patt 142[150:158:166], K2, m1, P1, rep from * once.
8th and 9th rounds * P1, K3, patt 142[150:158:166],

K3, P1, rep from * once.
Cont in this way keeping patt correct and inc 1 st at each side of seam sts on every 3rd round, until there are 13 sts on st st gusset section on each side of seam sts.
Work 2 rounds more or until work measures 16[16½:16½:17] in (41[42:42:43] cm) or required side length.

Divide for armholes
1st round P1, K13, slip these 14 sts and last 14 sts of previous round onto holder for sleeve, patt 142[150:158:166], turn and complete front on these sts, leaving rem sts on holder.
Cont in patt working in rows until front measures 23½[24¼:24¾:25½] in (60[62:63:65] cm), ending with a wrong side row.

Left shoulder strap
1st row (right side facing) (P2, K2) 12[13:14:15] times, turn.
Work strap on these sts leaving rem sts on holder.
2nd row (P2, K2) 12[13:14:15] times.
3rd and 4th rows (K2, P2) 12[13:14:15] times.
Rep these 4 rows 3 times more. Leave sts on holder.

Right shoulder strap
With right side of front facing slip centre 46 sts onto holder and leave for neckband. Rejoin yarn to rem 48[52:56:60] sts and work other strap in same way.

Back
With right side of rem sts facing, slip 2nd gusset sts onto needle and rejoin to rem 142[150:158:166] sts for back.
Work as for front including shoulder straps.

Join shoulders
With wrong side of sts for left front and back shoulder touching, cast off by knitting 1 st from both front and back needles tog twice, lift 1st st over 2nd in usual way, * knit tog next st from front and back, lift 1st over 2nd, rep from * until all sts have been cast off.
Work other shoulder in the same way.

Neckband
With set of 2¼mm needles and right side of work facing, rib across 46 sts from back holder, K up 24 sts along strap edge, rib across 46 sts from front holder and K up 24 sts along other strap edge. Cont in rounds keeping rib correct and dec as follows:
1st round * Rib 46, sl 1, K1, psso, K20, K2 tog, rep from * once.
Cont in this way dec 1 st at each side of neck gussets on every round until 2 sts rem.
Next round * K2, P2, rep from * to end.
Work 8 rounds more in rib. Cast off loosely in rib.

Sleeves
With set of 3mm needles and right side of right armhole facing, K up 94[98:102:106] sts round armhole, K13, P2, K13 from gusset.
1st round K2,* P2, K2, rep from * to last 28 sts, K11, K2 tog, P2, sl 1, K1, psso, K11.
2nd round K2, * P2, K2, rep from * to last 26 sts, K12, P2, K12.
3rd round P2, * K2, P2, rep from * to last 26 sts, K12, P2, K12.
4th round P2, * K2, P2, rep from * to last 26 sts, K10, K2 tog, P2, sl 1, K1, psso, K10.
Cont in this way, dec at each side of seam sts on every 3rd round but working basket stitch patt as set for 12 rounds more, then cont in st st for 9 rounds dec on 4th and 8th rounds.
P 4 rounds.
Cont in st st for 8 rounds, dec at each side of seam sts on every 4th round. P 2 rounds.
Cont in st st dec 1 st at each side of seam sts on every 6th round until 72[72:76:76] sts rem. Work until sleeve measures $15\frac{3}{4}[15\frac{3}{4}:17:18]$ in (40[40:43:46] cm) or $2\frac{1}{4}$ in (6 cm) less than required finished length.
Change to 2¼mm needles and work in K2, P2 rib for $2\frac{1}{4}$ in (6 cm).
Cast off.
Work other sleeve in same way, placing gusset at beg of round.

ERISKAY GUERNSEY
If pattern intrigues you, then a guernsey like this one will be a must. Because there is a logic about patterns, they are less difficult to knit than you may

153 *Eriskay guernsey*

think. In patterns like these the row or round you have just worked guides you on to the next row as the shapes are formed. Because it is patterned allover, only one size is given, but if you want another size, chart it out on graph paper making one section wider or smaller as required. The original pattern is over 50 years old and even the ribs between the patterns were patterned with a crossed stitch. The chart shows this and may be worked as an Indian corn stitch or by making a simple cable of one stitch over the other. If you prefer, it can be omitted without in any way spoiling the pattern.

Materials
14 50g balls of Poppleton's 4-ply Guernsey
1 set of 2¼mm (No. 13) needles or 1 set of 2¼mm needles and 1 circular 2¼mm needle
4 small buttons

Measurements
To fit 36–38in chest (92–97 cm)
Length to shoulder 25½ in (65 cm)
Sleeve seam 19 in (49 cm)

Tension
32 sts and 42 rows or rounds to 4 in (10 cm) over st st on 2¼mm needles
N.B. The 2 rib sts between panels may be worked throughout as Indian corn sts by making yon, K2, lift yon over both sts and off needle tip. They may also be worked as a crossed st if preferred by passing right needle behind 1st K st, knitting into back of 2nd st and then into front of 1st st and slipping both sts from needle together. Alternatively they may be left plain. If this crossed st is used, repeat every 8th row or round.

Instructions
With set of 2¼mm needles or circular needle cast on 288 sts.
1st round * P2, K2, rep from * to end.
Place marker thread before 1st st to mark beg of round.
Rep 1st round until rib measures 3 in (8 cm).
Next round P.
Next round K2, * K6, m1, K12, rep from * to last 16 sts, K6, m1, K10. 304 sts.
Next round P.

Begin pattern (154)
1st round * work from A to B of chart 7 times, then from A to C once, place marker thread before next st to show start of 3 seam sts, K1, m1 purlwise, K1, rep from * once. 306 sts.
2nd round * work from A to B of chart 7 times, then from A to C once, working 2nd row, K1, P1, K1, rep from * once.
Cont in this way until chart has been rep 3 times, keeping 3 seam sts as set.
Next round * (P2, K13, P2, K2) 7 times, P2, K13, P2, K1, P1, K1, rep from * once.
Next round * (P2, K13, P2, K2) 3 times, P2, K13, P2, K1, m1, K1, (P2, K13, P2, K2) 3 times, P2, K13, P2, K1, P1, K1, rep from * once. 308 sts.
This is approximately 3 in (8 cm) from completed side length. If longer garment is required, add additional rounds in pattern.

Begin border and gusset (155)
1st round * P151, K into front, back and front of next st, P1, K into front, back and **front of next** st, rep from * once.

2nd round * P151, K3, P1, K2, rep from * once.
3rd round * P1, work diamond pattern from chart 155, rep from A to B 18 times, then B to C once, (P1, K3) twice, rep from * once.
4th round as 3rd round, working 2nd row of chart.
5th round * P1, work diamond patt as set over 149 sts, P1, K1, m1, K2, P1, K2, m1, K1, rep from * once.
Cont in this way until diamond chart is complete, inc 1 st at each side of seam st, inside edge st, on every 4th round.
Next round * P1, K149, (P1, K9) twice, rep from * once.
Next round * P151, K1, m1, K8, P1, K8, m1, K1, rep from * once.
Next round * P151, K10, P1, K10, rep from * once, P 1st st of next round.

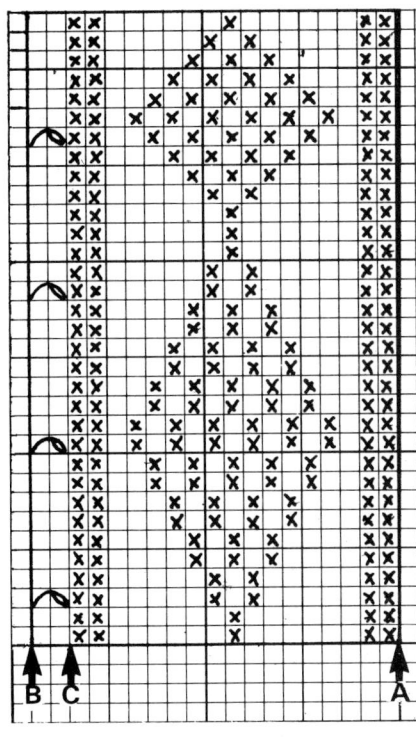

154 *Eriskay guernsey chart 1*

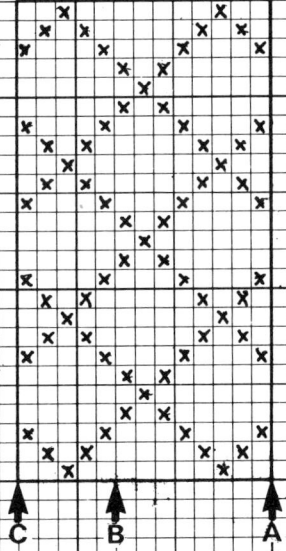

155 *Eriskay guernsey, chart 2*

Divide for armholes
1st row patt 149 sts working from chart 156 from A to dotted line, work st to left of dotted line and then complete from dotted line to A, noting that wave pattern does not reverse but faces the same direction in both sections. Slip both sets of 23 sts for gusset onto strand of yarn and leave rem 149 sts for back.

Front
Cont on 149 sts just worked, working in rows from chart until all rows have been worked, ending with a wrong side row. **

Divide for shoulder straps
*** *1st row* (right side facing) K50, turn, leaving rem sts unworked.
2nd row P.
3rd row P.
4th row K.
Rep these 4 rows 3 times, then 1st and 2nd rows once. Leave sts on holder. ***
With right side facing, slip centre 49 sts to holder for neckband.
Rejoin yarn to rem 50 sts at neck edge and complete as for other strap from *** to ***.

Back
With right side of rem 149 sts facing, rejoin yarn and work as for front to **.

To join shoulders
Place sts for shoulder of front and same number of sts at same edge for back together with right sides touching and cast off shoulders together on wrong side.

Neckband
With set of $2\frac{1}{4}$mm needles and right side of work facing, K up 8 sts from centre of shoulder strap to front neck edge, K across 49 sts from front holder, K up 17 sts from other strap edge, K across 49 sts from back holder, K up 8 sts from right half of strap, turn, cast on 5 sts.
Work in rows as follows:
1st row K5, P2, * K2, P2, rep from * to last 5 sts, K5.
2nd row K5, rib to last 5, K5.
3rd row as 1st.

156 *Eriskay guernsey, chart 3*

Buttonhole row K2, yon, K2 tog, K1, rib to last 5, K5. Cont as set working buttonholes on every 6th row. Work 3 rows after 4th buttonhole. Cast off.

Sleeves
With set of 2¼mm needles and right side of left armhole facing, K up 127 sts from gusset round armhole, P1, (K10, P1) twice across gusset.
1st round P to last 22 sts, (K10, P1) twice.
2nd round K to last 23 sts, P1, (K10, P1) twice.
3rd round as 1st.
Cont in st st, keeping seam st correct and dec 1 st at each side of gusset inside edge st every 4th round until all gusset sts have been worked off.
Cont dec every 6th round on each side of seam st until 90 sts rem. Work without shaping until sleeve measures 14 in (37 cm) or approximately 5 in (12 cm) less than required length.
Decrease round * K9, (P3 tog) twice, rep from * to end of round.

Begin cuff pattern
1st round * work 11 st pattern from chart 157, rep 6 times.
Rep this round until patt has been worked 3½ times. Cast off.
Work other sleeve in same way placing gusset at beg of round.

To complete
Sew four buttons onto garter st under flap after slip stitching edge to underside of buttonhole flap.

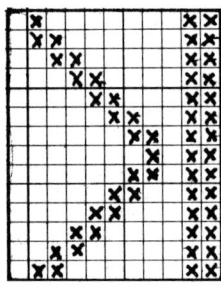
157 *Eriskay guernsey, chart 4*

POLPERRO GUERNSEY
Richard Searle's unusual patterned guernsey must have been knitted prior to 1860 for it was about 1860 when Lewis Harding started to photograph the 82 men of Polperro for his large composition which now hangs in the Rowett Institute in Polperro. It is a bold pattern, easy both to follow and to adjust to size if the sizes provided do not suit your needs. To increase the size the narrow ribs at each side could be made larger or the fern pattern could also be enlarged. To make this smaller work fewer stitches over the fern pattern, redrawing it on graph paper for easy working.

158 *Enlarged photograph of 130 year old gansey pattern worn by Richard Searle of Polperro Photo by Lewis Harding, c.1850*

Materials
15[16:17:18] 50g balls of Poppleton's 5-ply Guernsey
1 set of 2¾mm (No. 12) needles or 1 set of 2¾mm needles and one 2¾mm circular needle. (If short needles are being used for the body 2 sets may be required.)

Measurements
To fit chest 38[40:42:44] in (97[102:107:112] cm)
Length to shoulder $26\frac{3}{4}[27\frac{1}{2}:27\frac{3}{4}:28\frac{1}{2}]$ in (68[70:71:73] cm)
Sleeve seam 20[20:21:22] in (51[51:53:56] cm)

Tension
28 sts and 38 rows to 4 in (10 cm) over st st on $2\frac{3}{4}$mm needles

Instructions
Begin at lower edge.
With set of $2\frac{3}{4}$mm needles or circular needle cast on 264[280:296:312] sts.
1st round * P1, K2, P1, rep from * to end.
Place marker loops before 1st st and after 132nd[140th:148th:156th] st to mark sides.
Cont until rib measures 3 in (8 cm), inc 18 sts evenly on last round, 9 sts between 1st and 2nd marker and 9 sts between 2nd and 1st markers. 282[298:314:330] sts.

Place seam sts
1st round * P1, K139[147:155:163], P1, rep from * once.
Cont in this way keeping seam sts as set until work measures $12\frac{1}{2}[13:13\frac{1}{2}:14]$ in (32[33:33:34] cm) or approximately $3\frac{1}{2}$ in (9 cm) less than required side length.
For additional length add rows at this point.

Begin yoke
Next 2 rounds P

Begin gusset
1st round P1, slip this st onto end of last needle ready for gusset, * K139[147:155:163], P1, m1, P1, rep from * once.
2nd round * K139[147:155:163], P1, K1, P1, rep from * once.
3rd and 4th rounds * P140[148:156:164], K1, P141[149:157:165], K1, P1.
5th round * K139[147:155:163], P1, K into front, back and front of next st, P1, rep from * once.
6th round * K139[147:155:163], P1, K3, P1, rep from * once.
7th and 8th rounds P140[148:156:164], K3, P141[149:157:165], K3, P1.

Cont inc 1 st at each side of gusset before seam st on next and every 4th row AT SAME TIME begin yoke patt thus:

1st round * (K2, P1) twice, ** K2[3:4:5], work fern from chart 159 over next 17 sts, K2[3:4:5], P1, K2, P1, ** rep from ** to ** once, K2, work triangle patt from chart 160 over next 23 sts for centre panel, K2, *** P1, K2, P1, K2[3:4:5], work fern from chart 159 over next 17 sts, K2[3:4:5], *** rep from *** to *** once, (P1, K2) twice, work seam st, gusset sts and seam st, rep from * once.

Work 23 rounds more, inc gusset on every 4th row and rep fern and triangle patt from charts as required.

Divide for armholes
1st row K5, patt 129[137:145:153], K5, turn and complete front on these sts, leaving rem sts on holder.
Cont in patt, keeping 5 edge sts as K sts on every row, until front measures $25\frac{1}{4}$[26:$26\frac{1}{4}$:27] in (65[66:67:69] cm), ending with a wrong side row.

Divide for neck and shoulders
1st row patt 48[51:53:57], K2 tog, turn. Complete left shoulder on these 49[52:54:58] sts, leaving rem sts on holder.

Left side
Keeping patt correct, dec 1 st at neck edge on 2nd, 4th and 8th rows. Cont in patt until front measures $26\frac{3}{4}$[$27\frac{1}{2}$:$27\frac{3}{4}$:$28\frac{1}{2}$] in (68[70:71:73] cm), ending with a wrong side row. Leave rem 46[49:51:55] sts on holder.

Right side
With right side of front facing, slip centre 39[41:45:45] sts onto a holder and rejoin yarn at neck edge to rem 50[53:55:59] sts.
Keeping patt correct, dec 1 st at neck edge on 1st, 3rd, 5th and 9th rows. Work until same length as other side, ending with a wrong side row. Leave rem sts on holder.

Back
With right side of work facing, slip both gusset sts and seam sts at each side onto holder, leaving centre rem 139[147:155:163] sts.

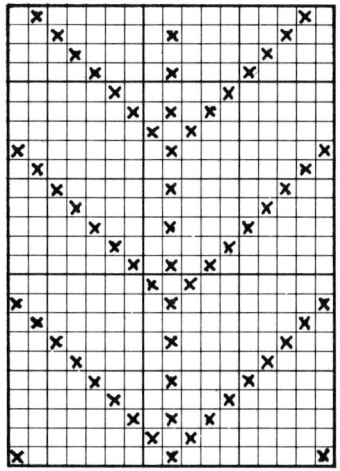

159 *Chart for Polperro gansey*

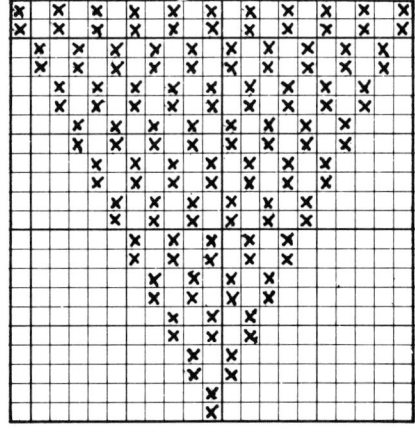

160 *Chart for Polperro gansey*

1st row (right side facing) K5, patt 129[137:145:153], K5.
Cont in patt keeping 5 K sts at edges on every row until work measures same length as front to shoulder, ending with a wrong side row.

Divide for shoulders
Divide sts on back into 3 groups 1 at each end for shoulder of 46[49:51:55] sts and a centre group for neck of 47[49:53:53] sts.
Place left shoulder on needle with point at armhole and place front left shoulder on needle with point at armhole.
Cast off both sets of sts in usual way on right side (wrong sides touching) by knitting tog one st from each side, lifting the sts one over the other every time 2 sts are on needle until all sts are worked.
Cast off other shoulder in same way.

Neckband
With set of $2\frac{3}{4}$mm needles and right side of work facing, K across 47[49:53:53] sts from back, K up 15 sts from side of neck K across centre 39[41:45:45] sts and K up 15 sts from other front side.
Work in rounds of K2, P2 rib for $1\frac{1}{2}$ in (4 cm). Cast off in rib.

Sleeves
With set of $2\frac{3}{4}$mm needles and right side of left armhole facing, rejoin yarn at start of gusset, P1, K15, P1, K up 105[109:117:121] sts.
1st round P1, K15, P1, K to end.
Rep 1st round 3 times.
5th round P1, sl 1, K1, psso, K11, K2 tog, P1, K to end.
6th round P1, K13, P1, K to end.
Rep last round twice.
9th round P1, sl 1, K1, psso, K9, K2 tog, P to end.
10th round P1, K11, P to end.

Place pattern
1st round P1, K11, P1, K38[39:42:43], P1, K2, P1, K2[3:4:5], work panel over next 17 sts working from top of chart A downwards, K2[3:4:5], P1, K2, P1, K to end.
Cont on these sts working patt as set until 6 chevrons have been completed, then work 2 rounds

P, *at same time* dec on gusset every 4th round until gusset sts are worked off, then every 6th round until 73[77:77:81] sts rem.
Work without shaping until sleeve measures 16[16:17:18] in (41[41:43:46] cm) or 4 in (10 cm) less than required length, dec 1 st on last round.
Work in K2, P2 rib for 4 in (10 cm). Cast off in rib.
Work other sleeve in same way placing gusset at end of round.

SHERINGHAM GANSEY

Worked in fine yarn, this Sheringham gansey was actually photographed in the same year as the gansey worn by Robert 'Tarr' Bishop in picture 47, and may well have been an experiment that led to the very well-designed pattern he is wearing. There are certainly errors in the gansey in this picture, but the following instructions will let you make one in the Sheringham style, with the errors removed. The pattern repeat is not quite as perfect as the other, more complex, pattern because the line does meet the diamond, losing a degree of clarity once every eight diamonds, but it is an easy pattern to knit. It is a knitter's pattern: in other words, the repeat tells you where you are all the time. Although the diamond repeats over ten rows, and the line over eight, it is not difficult to remember.

Materials
12[13:14] 50g balls of Poppleton's 4-ply Guernsey
1 set of $2\frac{1}{4}$mm (No. 13) needles or 1 set of $2\frac{1}{4}$mm needles and 1 circular $2\frac{1}{4}$mm needle

Measurements
To fit chest 36–38[40–42:44–46] in (92–97[102–107:112–117] cm)
Length to shoulder 25[$26\frac{1}{2}$:$27\frac{1}{2}$] in (64[67:70] cm)
Sleeve seam 18[20:21] in (46[51:54] cm)

Tension
32 sts and 42 rows to 4 in (10 cm) over st st on $2\frac{1}{4}$mm needles

Instructions
With set of $2\frac{1}{4}$mm needles or circular needle cast on 296[332:368] sts.
1st round * P1, K2, P1, rep from * to end.

161 *Sheringham fisherman*

Place marker thread before 1st st and after 148th[166th:184th] sts to mark side edges.
Cont in rib as set until work measures 2 in (5 cm), inc 18 sts evenly on last round, 9 sts between 1st and 2nd markers and 9 sts between 2nd marker and end of round. 314[350:386] sts.

Place seam sts
Increase round * m1, K157[175:193], rep from * once. 316[352:388] sts.
1st round * P1, K157[175:193], rep from * once.
2nd round K.
Rep last 2 rounds until work measures $11[11\frac{3}{4}:12\frac{1}{4}]$ in $(28[30:31]$ cm) or $4\frac{3}{4}$ in (12 cm) less than required side length, ending with 1st round.

Begin gusset and yoke
1st round * K into front, back and front of next st, P157[175:193], rep from * once.
2nd round * P1, K1, P158[176:194], rep from * once.
3rd round K.
4th round * P1, K1, P1, K157[175:193], rep from * once.
5th round * K1, (m1, K1) twice, P157[175:193], rep from * once.
6th round * P1, K3, P158[176:194], rep from * once.

162 *Chart for Sheringham pattern*

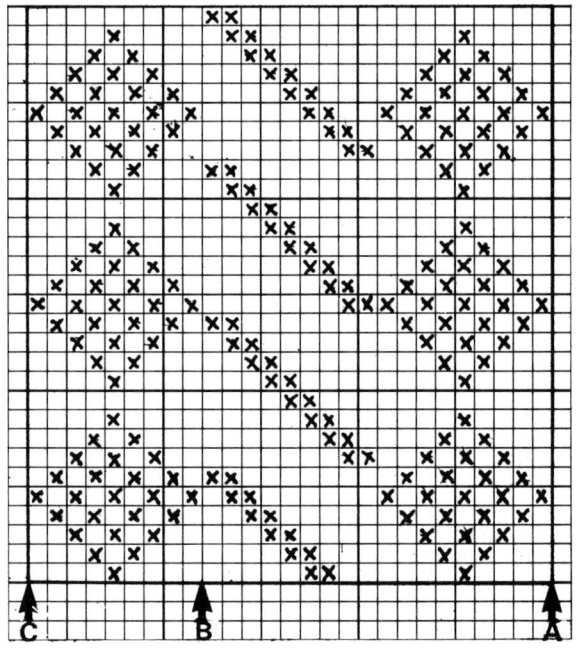

7th round K.
8th round * P1, K3, P1, K157[175:193], rep from * once.
9th round * K1, m1, K3, m1, K1, P157[175:193], rep from * once.
10th round * P1, K5, P1, K157[175:193], rep from * to end.

Begin yoke pattern
1st round * K7 for gusset, K2, rep from A to B on chart 8[9:10] times, work from B to C, K2, rep from * once.
2nd round * P1, K5, P1, K2, work patt over next 153[171:189], K2, rep from * once.
Cont in this way, inc 1 st at each side of both gussets on next and every 4th row until there are 27 sts on each gusset including P st at each side. AT SAME TIME rep diamond patt every 10th round and line pattern every 8th round.
Work 3 rounds after reaching 27 sts or until work measures $15\frac{3}{4}$[$16\frac{1}{2}$:17] in (40[42:43] cm).

Divide for armholes
Turn and work in rows.

Front
1st row (wrong side facing) P2, patt 153[171:189], P2, turn.
Complete front on these 157[175:193] sts, leaving rem sts on holder.
Cont until work measures 24[$25\frac{1}{2}$:$26\frac{1}{2}$] in (61[65:68] cm), ending with a wrong side row. **

Divide for shoulder straps
*** *1st row* P52[58:64], turn. Work left shoulder strap on these sts, leaving rem sts on holder.
2nd row K.
3rd row K.
4th row P.
Rep these 4 rows 4 times more then 1st and 2nd rows once.
Leave these sts on holder until back is worked. ***
With right side of front facing slip centre 53[59:65] sts onto holder and rejoin yarn at neck edge to rem 52[58:64] sts and work as for other strap from *** to ***

Back
Slip both sets of 27 sts for gussets onto strands of yarn or holders and with wrong side facing rejoin yarn to rem 157[175:193] sts.
Work as for front to **.

Join shoulders
Place front and back left shoulders together with needle points at armhole edge, and cast shoulders off on wrong side working sts tog. Work other shoulder in same way leaving centre sts on holder.

Neckband
With set of $2\frac{1}{4}$mm needles and right side facing, K across 53[59:65] sts from back holder, K up 15 sts along neck edge of strap, K across 53[59:65] from front holder and K up 15 sts from other strap.
Work in rounds of K2, P2 rib for $1\frac{1}{2}$ in (4 cm). Cast off.

Sleeves
With set of $2\frac{1}{4}$mm needles and right side of work facing K up 121[137:153] sts from gusset round left armhole, (P1, K12) twice, P1.
Keeping seam sts correct and dec 1 st at each side of gusset on every 4th round, on K rounds, (P2 rounds, K2 rounds) twice, P2 rounds.

Begin pattern
1st round K2[1:0], work from A to B of patt until 11[10:9] sts from seam st, work from B to C for 9 sts, K2[1:0], work gusset sts. Cont in patt until 4 diamonds have been completed AT SAME TIME dec as before until all gusset sts are worked off then dec at each side of single seam st.
Keeping dec and seam st correct, (P2 rounds, K2 rounds) twice, P2 rounds. Cont in st st, keeping seam st correct and dec at each side of seam st on every 6th round until there are 73[81:85] sts. Work without shaping until sleeve measures 16[18:19] in (41[46:49] cm) or 2 in (5 cm) less than required sleeve length, dec 1 st in centre of last round.
Work in rounds of K2, P2 rib for 2 in (5 cm). Cast off.

163 *Skipper Donald Thomson, Thurso*

THURSO GANSEY
Knitted around 1910, this gansey, worn by Donald Thomson, skipper of *The Children's Friend*, is a pattern which every knitter finds adaptable to all sizes. It is an interesting pattern to knit and gives a variety of textures with its ropes, chevrons and diamonds. The instructions are for a single moss diamond but if preferred you can work these as purl, which is a little more unusual. Alternatively, you could give it your own variation by alternating moss and purl diamonds, one above the other, or using one type in the centre and the other at the sides.

Materials
16[17:17:18] 50g balls of Poppleton's 5-ply Guernsey
1 set of $2\frac{3}{4}$mm (No. 12) needles or 1 set and 1 circular $2\frac{3}{4}$mm needle for body. (If short double pointed needles are used 2 sets may be required to hold the total number of stitches with ease)
1 cable needle
3 buttons

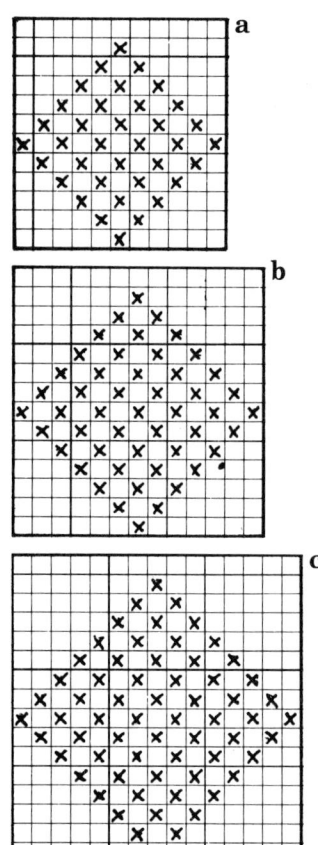

164 *Charts for diamonds*

Measurements
To fit 38[40:42:44] in chest (97[102:107:112] cm)
Length to shoulder $25\frac{1}{2}$[$26\frac{1}{4}$:$26\frac{1}{4}$:27] in (65[67:67:69] cm)
Sleeve seam 20[20:21:22] in (51[51:53:56] cm)

Tension
28 sts and 38 rows or rounds to 4 in (10 cm) over st st on $2\frac{3}{4}$mm needles

Special abbreviation
C6 sl next 3 sts to CN, hold at back of work, K next 3 sts, K3 sts from CN.

Instructions
Begin at lower edge of body.
With $2\frac{3}{4}$mm double-pointed needles or $2\frac{3}{4}$mm circular needle cast on 280[296:312:328] sts.
1st round * P1, K2, P1, rep from * to end.
Place a coloured marker loop of thread before 1st st to mark beg of round and after 140th[148th:156th:164th] st to mark other side. Rep 1st round until rib measures 3 in (8 cm), inc 11[9:9:7] sts evenly between 1st and 2nd markers and 11[9:9:7] sts between 2nd marker and end of round on last row. 302[314:330:342] sts.

Place seam sts and beg st st
1st round * P1, K149[155:163:169], P1, rep from * once.
Cont in st st, keeping 2P seams sts at each side until work measures $11\frac{1}{2}$[$11\frac{3}{4}$:12:$12\frac{1}{2}$] in (29[30:31:33] cm) or approximately 5 in (12 cm) less than required side length. For additional length add required number of rounds at this point.

Work ridges
Next 3 rounds P.
Next 2 rounds * P1, K149[155:163:169], P1, rep from * once.
Next 3 rounds P.
Next round * P1, K149[155:163:169], P1, rep from * once.

Begin patt and gusset
1st round * P1, m1, P3, ** over next 11[13:13:15] sts work 1st row of diamond from chart 164a[b:b:c], P2, (K6, P2) twice, over next 19[19:23:23] sts work 1st row of chevron from chart 165a[a:a:b], P2, (K6, P2) twice, ** rep from ** to ** once, over next 11[13:13:15] sts,

work 1st row of diamond from chart 164a[b:b:c], P3, m1, P1, * rep from * to * once.

2nd round * P1, K4, ** work 2nd row of diamond, K18, work 2nd row of chevron, K18, ** rep from ** to ** once. Work 2nd row of diamond, K4, P1, * rep from * to * once.

3rd round * P1, K1, P3, ** work 3rd row of diamond, P2, (K6, P2) twice, work 3rd row of chevron, P2, (K6, P2) twice, **, rep from ** to ** once, work 3rd row of diamond, P3, K1, P1, * rep from * to * once.

4th round * P1, K4, ** work 4th row of diamond, K18, work 4th row of chevron, K18, ** rep from ** to ** once, work diamond, K4, P1, *, rep from * to * once.

5th round * P1, m1, K1, P3, ** work diamond, P2, (C6, P2) twice, work chevron, P2, (C6, P2) twice, ** rep from ** to **, work diamond, P3, K1, m1, P1, * rep from * to * once.

Cont in this way rep diamond and chevron patts as required and inc 1 st inside seams sts on every 4th round, working the gusset sts in st st throughout. Cont until there are 26 sts on gusset including 2 seams sts. Work 3 rounds more or until work required length to underarm, 16[16½:16¾:17¼] in (41[42:43:44] cm), ending last round *before* seam st.

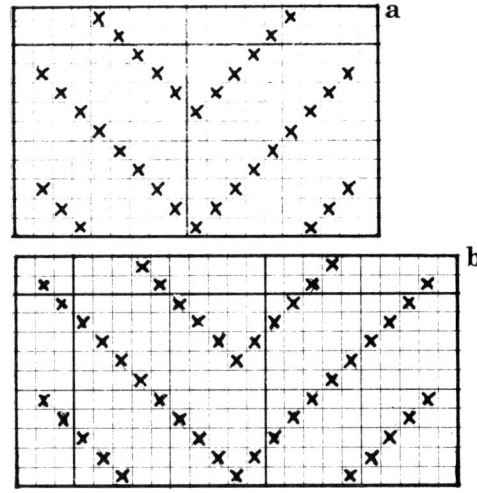

165 *Charts for chevrons*

Divide for armholes
Slip 12 gusset sts both seam sts and 12 gusset sts from start of next needle onto strand of yarn until required. Turn and with wrong side facing patt 149[155:163:169] sts, turn. Complete back on these sts, leaving rem sts on holder.
Cont in patt as set, working in rows until back measures 24½[25¼:25¼:26] in (63[65:65:67] cm) or 1 in (2.5 cm) less than required length to shoulder, ending with a wrong side row. Do NOT break off yarn but leave sts on holder.
With wrong side of work facing slip 26 sts for 2nd gusset onto thread or holder and rejoin yarn to rem 149[155:163:169] sts.
Work front in rows until same length as back, ending with a wrong side row. Do NOT break off yarn.

Divide for shoulders
Divide sts of front onto 3 needles, 2 groups of 49[51:54:56] sts at each end and a centre group of 51[53:55:57] sts for neck.
Divide back in same way.

Shoulder strap
Slip sts for both back and front right shoulder onto needles with points at armhole edge. Take needle with sts for front in right hand and insert needle tip between 1st and 2nd sts of back with right side of back facing, lift yarn at this edge and cast on 15 sts to lie between back and front sts. Do not turn work.
1st row (right side facing) K14 of 15 sts cast on, sl last st knitwise, K1 st from back, psso, turn.
2nd row Sl 1 knitwise, P13, P tog last st with 1st st from front.
3rd row Sl 1 knitwise, P1, (K5, P1) twice, P1, sl last st knitwise, K next st from back, psso, turn.
4th row Sl 1 knitwise, P5, K1, P1, K1, P5, P tog last st with next st from front, turn.
5th row Sl 1 knitwise, K4, P1, K3, P1, K4, sl last st knitwise, K next st from back, psso, turn.
6th row Sl 1 knitwise, P3, K1, P5, K1, P3, P tog last st with next st from front, turn.
7th row Sl 1 knitwise, K2, P1, K7, P1, K2, sl last st knitwise, K next st from back, psso, turn.
8th row Sl 1 knitwise, P1, K1, P9, K1, P1, P tog last st with next st from front, turn.
Rep from 3rd–8th rows until all shoulder sts in these groups are worked off. Leave rem sts on holder.
Work other shoulder in the same way, reading front for back and back for front. Leave sts on holder.

Neckband
With right side of left shoulder facing, slip 10 sts at right of shoulder strap onto holder.
With set of 2¾mm needles or circular needle work sts round neck thus:
Cast on 4 sts and place on right needle, with same needle K5 sts rem on left shoulder strap, K51[53:55:57] sts from front K15 sts from right strap, K51[53:55:57] sts from back, and K10 sts from shoulder strap sts on holder. 136[140:144:148] sts.
1st row K7, * P2, K2, rep from * to last 5 sts, K5.
2nd row K5, * P2, K2, rep from * to last 7 sts, P2, K5.
3rd row as 1st row.
4th row K5, * P2, K2, rep from * to last 7 sts, P2, K2, yon, K2 tog, K1.
Keeping 5 garter st border at each end correct work 5 rows more.
Rep last 6 rows twice more.
Cast off in rib.

Sleeves
With set of 2¾mm needles and right side of left armhole facing, rejoin yarn at start of gusset, thus: K12 from gusset, P2 seam sts, K12 from other side of gusset, K up 99[103:107:111] sts.
Next round K12, P2, K to end.
P 2 rounds

Begin patt band
1st round K12, P2, K12, K1[3:1:3], * P2, K5, P1, rep from * to last 2[4:2:4] sts, P1, K1[3:1:3].
2nd round K12, P1, K12, K1[3:1:3], * P1, K1, P1, K3, P1, K1, rep from * to last 2[4:2:4] sts, P1, K1[3:1:3].
3rd round K10, K2 tog, P2, sl 1, K1, psso, K10, K1[3:1:3], * P1, K2, P1, K1, P1, K2, rep from * to last 2[4:2:4] sts, P1, K1[3:1:3].
4th round K11, P2, K11, K1[3:1:3], * P1, K3, rep from * to last 2[4:2:4] sts, P1, K1[3:1:3].
5th round K11, P2, K11, K1[3:1:3], * P1, K7, rep from * to last 2[4:2:4], P1, K1[3:1:3].
6th round work as for 5th round.
Cont in this way until 7 chevrons have been completed, dec 1 st at each side of gusset on every 4th round.
Next 2 rounds P.
Cont in st st, keeping 2 seam sts correct and dec 1 st at each side of seam sts on every 6th round once gusset sts have been completed. Work until 85[89:93:97] sts rem then dec every 8th round until 73[77:77:81] sts rem.
Work without shaping until sleeve measures 17[17:18:19] in (43[43:46:49] cm) or 3 in (8 cm) less than required finished length.
Work 3 in (8 cm) in K2, P2 rib, dec 1 st at beg of 1st round. Cast off in rib.
Work other sleeve in same way placing gusset at end of round.

To complete
Sew cast on 4 st flap over underflap at neck edge.
Sew 4 buttons on underflap to match buttonholes.

Bibliography

ANSON, P. *Fishing Boats and Fisher Folk on the East Coast of Scotland* The Faith Press, London, 1930
— *Fishermen and Fishing Ways* The Faith Press, London, 1931
— *Scots Fisher-Folk* The Faith Press, London, 1953
de BURLET, S. *Port of Polperro* Rooster, Falmouth, 1977
COUCH, J. *History of Polperro* (manuscript) 1871
COURTNEY, M.A. *Cornish Feasts and Folklore* Beare, Penzance, 1890
CRUPPLES, Mrs. *Newhaven: Its Origins and History* Edinburgh, 1888
GOURLAY, G. *Memories of Cellardyke* Cupar, 1879
HARTLEY, M. and INGILBY, J. *Old Handknitters of the Dales* Dalesman Press, Yorkshire, 1951
LANYON, A. *The Rooks of Trelawne* Photographer's Gallery, London, 1976
MACDONALD, J. and GORDON, A. *Down to the Sea* Shandwick, 1971
MCIVER, D. *An Old Time Fishing Town, Eyemouth* John Menzies, Glasgow and Edinburgh, 1906
NORBURY, J. *Odhams Encyclopedia of Knitting* Odhams, London
PAYNTER, S.W. *Old St Ives*
PEARSON, M. *Traditional Knitting Patterns: The Fisher Ganseys of North East England* Esteem Press, Newcastle upon Tyne, 1980
SLATER, J. *A Sea-faring Saga* Commercial Fishing Enterprises, Fleetwood, Lancs. 1979
THOMPSON, G. *Guernsey and Jersey Patterns* Batsford, London, 1955
WRIGHT, M. *Cornish Guernseys and Knit-frocks* Ethnographica, London, 1979
UTTLEY, J. *The Story of The Channel Islands* Faber and Faber, London, 1966

Conversion charts

CONVERSION CHART FOR BALL WEIGHTS

Balls vary in size and may be marked in ounces or grams. This chart shows the number of 20g, 25g and 50g balls required when the original amount is shown in ounces.

oz	20g	25g	50g	oz	20g	25g	50g
1	2	2	1	13	19	15	8
2	3	3	2	14	20	16	8
3	5	4	2	15	22	17	9
4	6	5	3	16	23	19	10
5	8	6	3	17	25	20	10
6	9	7	4	18	26	21	11
7	10	8	4	19	27	22	11
8	12	10	5	20	29	23	12
9	13	11	6	21	30	24	12
10	15	12	6	22	32	25	13
11	16	13	7	23	33	27	14
12	17	14	7	24	34	28	14

NEEDLE SIZE CHART

Metric (cm)	English before metrication	American
2	14	0
$2\frac{1}{4}$	13	1
$2\frac{1}{2}$		
$2\frac{3}{4}$	12	2
3	11	
$3\frac{1}{4}$	10	3
$3\frac{1}{2}$		4
$3\frac{3}{4}$	9	5
4	8	6
$4\frac{1}{2}$	7	7
5	6	8
$5\frac{1}{2}$	5	9
6	4	10
$6\frac{1}{2}$	3	$10\frac{1}{2}$
7	2	
$7\frac{1}{2}$	1	
8	0	11
9	00	13
10	000	15

METRIC CONVERSION CHART

cm	in
0.5	$\frac{3}{16}$
1	$\frac{3}{8}$
2	$\frac{3}{4}$
3	$1\frac{1}{8}$
4	$1\frac{5}{8}$
5	2
6	$2\frac{3}{8}$
7	$2\frac{3}{4}$
8	$3\frac{1}{8}$
9	$3\frac{1}{2}$
10	4
20	$7\frac{7}{8}$
30	$11\frac{7}{8}$
40	$15\frac{3}{4}$
50	$19\frac{5}{8}$
100	$39\frac{3}{8}$

GLOSSARY OF ENGLISH AND AMERICAN TERMS

English	American
Cable needle (CN)	Double-pointed needle (dpn)
Cast off	Bind off
Double crochet (dc)	Single crochet
Treble	Double crochet
Double treble	Treble
Stocking stitch	Stockinette
Tension	Gauge

List of Suppliers

Yarn
Richard Poppleton and Sons, Ltd
Albert Mills
Horbury
Wakefield
West Yorkshire
WF4 5NJ

(0924 – 272376)

In case of difficulty in obtaining Poppleton's 5-ply Guernsey, please write to the above address for stockists' list or to arrange for the purchase of their yarn.

Needles
Stove and Smith
98 Commercial Street
Lerwick
Shetland

Steel double-pointed needles are still available from this address and were the type of 'wire' used throughout Scotland for guernsey knitting.

Leather knitting pads
Goodlad and Goodlad
90 Commercial Street
Lerwick
Shetland

Leather knitting pads (tippees or wiskers) are available from the above address.

Index

Abb wool 2, 40
Aberdeenshire 86
Allover diamond pattern chart 66
Alternating moss and purl diamond pattern chart 66
'Amazons' 75
Anstruther 73, 80
Appledore 29, 38; jersey instructions 104; jersey shoulder strap 106
Arbroath diagonal line pattern chart 84
Arbuthnot Museum 84
Armholes 19

Banded patterns 14, 15, 16, 17, 38, 80; pattern charts 80
Basket stitch 16, 18, 38, 101
'Betty Martin' pattern 18, 55, 63
Bishop, Robert 'Tarr' 41; pattern chart 42
Birds' e'en stitch 13
Block patterns 13
Bothy blankets 82
Box stitch 18, 101
Braces 35
Bridlington 56; gansey instructions 109; instruction charts 111
Brixham 40
Brooks 80, 83
Buckhaven 73
Bye-names 89

Caister 41, 47, 73
Caithness 13; flag pattern chart 93; shoulder strap 30
Campbeltown guernsey instructions 114
Carr, Mrs Pheobe 52; Mrs Carr's pattern chart 52
Casting off together 28
Casting on, Channel Island 23; knotted 22; three strand 22; thumb method 21; two needle method 21
Cat's teeth 13, 89
Cellardyke 70, 83
Census, 1851 35
Channel Island guernseys 19; knitting trade 20
Charts 10; using 100
Chevron, rope and moss pattern chart 59
Chevron pattern 35
Church windows pattern and chart 85
Coil of rope 41, 44, 47
Concertina player's pattern chart 35
Contract knitters 31
Cooper, Jim 'Coalie', pattern and chart 45, 46
Cooper, 'Cutty's pattern and chart 44, 45; young 'Cutty's pattern chart 44
Cornish Guernseys and Knit-Frocks 31
Cornish knitting 18, 40, 92; lattice pattern chart 38
Cornwall 14, 31, 41, 57, 89, 90
Couch, Dr Jonathan 32, 33
Cox, John 'Snouts' 16, 48; pattern and background pattern chart 48
Crail 73
Crimlisk, Mr J. 61
Cromer 50
Curtis, Jim, guernsey 35; diamond pattern chart 37

Devon 38
Diagrams 99
Diagonal line and garter stitch pattern chart 83
Diamond border chart 37
Diamond in box pattern chart 75
Diamond and rope pattern chart 65
Diamonds, single moss 16; double moss 16
Double moss stitch 6
Dumble, Jim, pattern and pattern chart 45, 46

Eriskay 92, 94; guernsey instructions 118; instruction charts 1 and 2 121; instruction chart 3 122; instruction chart 4 124; starfish pattern chart 96
Eyemouth 70, 75; disaster 73

Fair Isle knitting 19
Fern panel pattern and chart 67, 68
Fife 70, 76, 83
Filey 61, 92; fishermen's choir 63
Firth of Forth 73
Fishermen's Walk 73
Fisherrow 73; pattern 75
Flag and chevron pattern chart 78
Flag and rib pattern chart 78
Flamborough Head 13, 44, 52, 55; traditional pattern 58, 75; moss pattern chart 53
Foula 43
Freeman, Harry 15, 101
Front-cover charts 94, 95

Garter stitch and rib pattern chart 62
Giles, Mr John, returning officer 35
Grafting 29, 30
Graph paper 99
Guernsey and Jersey Patterns 8, 42, 61
Guernsey, origin 19; name 20
Guidance chart 102
Gushet 26
Gusset 26; outlined with single stitch 26; with centre seam stitch 27; northern gusset 27

Half diamond and rope pattern chart 91
Hailstones 13, 41, 44
Harding, Mr Lewis 33
Harvey, Michael 42
Heart and rope pattern chart 90
Hebridean shoulder strap 30
Herringbone pattern chart 85
Herring fishing 73
Hopeman 86

Indian corn stitch 29; pattern 96
Initials 59, 60

Jersey-frock 20, 89
Joliffe, Charles Jnr. 36

Kerr, Mrs, pattern and chart 48, 50
Knaggs, John 53, 54; pattern chart 55
Knit-frock 20, 32
Knitting pad 62; fish 57; sheaths 57; sticks 31, 40, 57

Laidlaw, Mrs, pattern and chart 70
Laidlaw's of Keith 82
Lansallos 33
Lanyon, Andrew 35
Line and block pattern chart 45
Liskeard 31, 37
Little, Henry Valentine 'Pinny' 47; pattern and chart 47
Lizard, the 38
Looe 31, 37
Lossiemouth 89
Lower Bodham 42
Lowestoft 41, 73

Mackay, Angus, pattern and chart 92
Mainprize, George, pattern and chart 56
Map, location 7
Marriage 14
Mary-Ann's stitch 18, 86, 101
Mayes, Billy 'Clubfoot' 43
Morayshire 26, 84, 85, 86
More, Mrs Mary 86
Morwenstow, Vicar of, pattern chart 37
Musselburgh 73

Necklines, working of 30; buttoned 30
Needles 20, 41, 51
Newhaven 15; banded patterns 15
Norfolk 13, 48, 90
Northcott, John, pattern chart 37

Nurse, Mrs Esther, pattern and chart 42

Old Flamborough pattern chart 55
Outer Hebrides 29

Pashby, Lizzie Ann 61
Patterns, block 37; brocade 19, 41; cabled 38, 41, 48; recording 16; wealth of 70
Paynter, S.W. 57
Peterhead 25
Plymouth 14
Polperro 32, 92; portrait of fishermen 33
Poppleton's yarn 63, 82, 100
Portsoy 87
Print o' the hoof and diamond pattern chart 64

Reas, Skipper John, pattern chart 71
Regional differences 14; similarities 14
Reid, George, pattern and chart 93
Rig' and furrow shoulder 29
Road to Duffus pattern
Robin Hood's Bay 8, 67, 100
Rook, Gilbert 'Leather' 16
Rooks of Trelawne 35
Rope and 'Betty Martin' pattern chart 56
Rope and seed pattern chart 65
Rope and step pattern chart 64
Rope, basket and open diamond pattern chart 58
Rope, 'Betty Martin' and ladder pattern chart 64
Rowett Institute 33, 124
Royal Highland Show 70

St Andrews 7
St Ives 38; history of 57
St Monance 73
Scales, Mrs, 53, 63
Scilly Isles 38

143

Scotland 13, 29, 30, 56, 69
Scottish Fisheries Museum 83
Scottish patterns 14; influence 47; flag pattern and chart 76
Seahouses 70
Seamen's Mission 15
Seam stitches 25; single purled 25; double seam stitches 25, 26; decorative seam stitches 25, 26
Searle, Richard 35, 42, 68; gansey instructions 125; instruction charts 127
Seed stitch patterns 18
Sennen Cove 38
Shepley, Mr 53
Sheringham 14, 15, 41, 44; gansey instructions 129; instruction chart 129
Shetland 19, 43, 96
Shoulders, cast off together 27; continuation of front 29; rig' and furrow 29;

Caithness shoulder strap 30
Single line wave pattern chart 83
Single moss stitch 6
Sizes 103
Slater, Mr James 90
Snake cable and ladder pattern and chart 38, 39
Stewart, Mrs Isabella 89
Stitch names 13
Sutherland, Mrs 89

'Tanker's guernsey 84; pattern chart 85
Tension 99
Thompson, Gladys 8, 42, 61
Thomson, Donald, pattern and chart 92, 93
Thurso, 14, 30, 91; gansey instructions 133; instruction charts 134, 135
'Tippee' 63
Tips 103
Traditional Flamborough pattern 58
Trails 38
Travel 14
Tree and rope pattern chart 85
Tree and twin rope pattern 91

Walkington, Fred and Carol 58, 59, 109; Grant and Craig 110
Welts, double 59; plain 24, 25; ribbed 25
Whalsay 26, 97
Whitby 1, 2, 15
Wick 89
Wires 20, 89
Wisker 63
Wright, Mary 31

Yarmouth 14, 41, 73
Yorkshire 17, 41
Yorkshire patterns, 14, 40

Zigzag line pattern chart 82